Introduction to Medical Geology

Focus on tropical environments

Erlangen Earth Conference Series

Series Editor

André Freiwald
University Erlangen, Germany

For further volumes:
http://www.springer.com/series/7037

C.B. Dissanayake · Rohana Chandrajith

Introduction to Medical Geology

Focus on tropical environments

Prof. Dr. C. B. Dissanayake
University of Peradeniya
Dept. Geology
Peradeniya
Sri Lanka
cbdissa@hotmail.com

Dr. Rohana Chandrajith
University of Peradeniya
Dept. Geology
Peradeniya
Sri Lanka
rohanac@hotmail.com

ISBN 978-3-642-00484-1   e-ISBN 978-3-642-00485-8
DOI 10.1007/978-3-642-00485-8
Springer Dordrecht Heidelberg London New York

Library of Congress Control Number: 2009926840

© Springer-Verlag Berlin Heidelberg 2009

This work is subject to copyright. All rights are reserved, whether the whole or part of the material is concerned, specifically the rights of translation, reprinting, reuse of illustrations, recitation, broadcasting, reproduction on microfilm or in any other way, and storage in data banks. Duplication of this publication or parts thereof is permitted only under the provisions of the German Copyright Law of September 9, 1965, in its current version, and permission for use must always be obtained from Springer. Violations are liable to prosecution under the German Copyright Law.

The use of general descriptive names, registered names, trademarks, etc. in this publication does not imply, even in the absence of a specific statement, that such names are exempt from the relevant protective laws and regulations and therefore free for general use.

Cover design: WMXDesign GmbH, Heidelberg

Printed on acid-free paper

Springer is part of Springer Science+Business Media (www.springer.com)

Dedicated to the

**ALEXANDER VON HUMBOLDT
STIFTUNG/FOUNDATION**

&

to

Professor Dr. Heinz J. Tobschall
Chair of Applied Geology, Friedrich-Alexander University of
Erlangen-Nürnberg, Germany

Foreword

Tropical lands are unique in a number of ways. Over150 countries have at least half their landmass in the tropics and these represent more than 40% of the world's population of around 6 billion. Only two regions namely, Singapore and Hong Kong lying in the tropical zone, rank among the 30 countries classified as those with high income by the World Bank. The geography, geology, extreme rock weathering, depletion of essential elements, agriculture, biodiversity and ecosystems, poverty, among others all contribute to the uniqueness of these land masses. Most interestingly, the fact that the vast majority of these people live in close association with the geological environment points to a remarkable association with certain diseases. This fact is highlighted by the authors in a number of interesting case studies involving millions of people. The role of certain trace elements, their geochemical pathways under the extreme climatic conditions as seen in the tropical environments lead to a fascinating aetiology of certain tropical diseases as exemplified by dental and skeletal fluorosis, iodine deficiency disorders and diseases caused by a lack of essential trace elements. This would almost certainly create a major interest among the medical fraternity on the importance of natural geological and geochemical processes and pathways in the aetiology of some tropical diseases.

A central premise of this book is that humans are also well and truly a part of the entire global environmental system and that they too are subjected to all the complex geochemical processes that operate around them. Logically therefore, humans who live in close contact with the immediate physical environment should also be influenced to a marked degree by the geochemistry of rocks, soils and water around them. It is this influence that has resulted in the emergence of the new discipline of Medical Geology. Nowhere is this better observed than in the tropical countries of the world.

The authors of this book have carried out significant research on the aspects mentioned above and their work has already been highlighted in several prestigious scientific journals.

This work can be considered as one of the pioneering efforts in the emerging discipline of Medical Geology, particularly on the tropical environment. There is clearly a dire need for the better understanding of the role of the geological processes in human health and this book fulfils a long felt global need.

Professor Dr. André Freiwald
Series Editor/ Erlangen Earth Conference Series
Erlangen, Germany

Preface

The emerging scientific discipline of Medical Geology has fascinated both scientists and laymen alike in view of its highly interdisciplinary nature. Medicine and Geology indeed form a truly awe inspiring scientific combination. The geosphere-biosphere interactions form an integral part of Medical Geology and some of its impacts cover millions of people the world over. From among these, those living in the tropical environments are particularly vulnerable to the effects of "geo-bio" relationships.

Tropical environments are unique from the point of view of their climate, soil characteristics, trace element deficiencies and enrichment, mineral imbalances, extreme cases of rock weathering and leaching out of nutrients, agricultural calamities among others. Millions of people living in these geologically and geochemically unique environments serve as classic examples of human beings who live in intimate contact with the geoenvironment and whose general health characteristics are markedly influenced by the geochemistry of the rocks, soils, water and plants found in their habitats. The influence of geochemistry on human and animal health is therefore best seen in the tropical environment. Medical Geology, as a scientific discipline has derived immensely from the earlier studies of the lands of the tropical belt of the world.

The extreme scarcity of text books covering the subject of Medical Geology is understandable in view of its very recent origin as a scientific discipline. It is the aim of this book to introduce the subject to students and researchers both in the fields of Medicine and Geology interested in geosphere-human interactions. It is clear that the geosphere does influence human health to a marked degree and it is up to the scientists to track the pathways of the disease - causing substances and elements originating from the geosphere and how they cross the biological barriers and enter the human body.

This book illustrates some interesting case studies of such geo-bio interactions affecting a very large population of the world. Throughout the book, the focus is on tropical environments and the impact of Medical Geology on millions of people, the vast majority of them living in developing countries of the tropical belt.

It should be emphasized that this book deals only with the relationship between natural geological factors and health in man and animals. As an introductory text it is not intended to deal with the medical aspects of the diseases mentioned in any great detail.

The authors wish to express their deepest gratitude to the Alexander von Humboldt Foundation of Germany, which funded the research that culminated in the publication of the book. Grateful thanks are due to Hema and Peter Dietze of Erlangen and also to Dr. med. Jayasumana and Nayananjali Jagoda, Erlangen, for providing accommodation during the time, in which this book was written. We thank colleagues and students of the Department of Geology of the University of Peradeniya, Sri Lanka and Institute of Geology and Mineralogy, University of Erlangen-Nürnberg for their assistance in various ways. We particularly would like to thank Mr. Rasika Mallawarachchi for typing the entire manuscript and Miss. Kushani Mahatantila for editorial assistance.

The authors wish to place on record the most valuable support and encouragement given by Professor Dr. Heinz J. Tobschall, Chair of Applied Geology, University of Erlangen-Nürnberg throughout the compilation of this book. All laboratory and office facilities at the Institute for Geology and Mineralogy granted by him to the authors are deeply appreciated. Further, his comments and criticisms have been most helpful to the authors in improving the scientific value of the book. The authors also wish to record their most sincere appreciation to Mr. Bill Campbell for his editorial corrections which undoubtedly improved the grammar and style.

Special thanks are due to Professor Dr. André Freiwald, the Series Editor of Erlangen Earth Conference Series for his guidance and support. We also thank Dr. med. Alexander Woywodt of for making extremely valuable comments on the medical aspects of the book. Finally, Dr. Christian Witschel and Mrs. Christine Adolph from Springer are thanked for undertaking the publication of this book.

C.B.Dissanayake
Rohana Chandrajith
Peradeniya, Sri Lanka
June, 2008

Acknowledgements

We are grateful to individual authors and following organizations who have kindly given permission for the reproduction of copyright material (figure number in parentheses).

American Association for the Advancement of Science (Figs. 7.9 & 7.10)
American Geophysical Union (Fig. 2.14)
American Medical Association (Fig. 10.4)
BGR, Germany (Fig. 6.9)
British Geological Survey (BGS) Permit Number IPR/88-20CGC[1]
    (Figs. 1.2; 4.17; 5.6; 5.9; 5.10; 6.6; 6.9; 7.15 & 7.16)
Center for Health and Population Research, Bangladesh (Fig. 7.17)
Current Science- India (Fig. 6.2)
Elsevier, the Netherlands (Figs. 2.2; 2.4; 4.3; 4.4; 5.11; 5.12; 5.14; 5.15; 5.17; 6.7; 6.12; 7.4; 7.5; 7.8; 7.13; 7.14; 8.3; 9.6; 10.3; 10.6 & 11.1)
E.Schweizerbart'sche Verlagsbuchhandlung OHG, Germany (Fig. 8.6)
Geological Society of France (Fig. 2.15)
Geological Society of India (Fig. 4.1)
International Fertilizer Industry Association (IFA) (Figs. 6.3 & 6.4)
S. Karger AG, Switzerland (Fig. 8.2)
Science Reviews Ltd., UK (Figs. 4.18 & 4.19)
Soil Science Society of America (Figs. 7.6 & 9.2)
Springer, Germany (Figs. 1.3; 2.12; 2.13; 5.5; 5.8; 9.1; 9.4 & 10.5)
Taylor and Francis-Balkema, the Netherlands (Figs. 2.8 & 2.9)
Taylor and Francis-UK (Figs. 4.13; 5.3 and 9.5)
The Geological Society, London (Figs.1.4; 2.5; 2.10; 4.1 & 5.4)
University of California, ANR Communication Services (Fig. 6.5)
United State Geological Survey (Fig. 4.16)
Wiley-VCH GmbH & Co KGaA, Germany (Figs. 3.1 & 9.3)

[1]The above figures were produced by the British Geological Survey; British Geological Survey and the Department of Public Health Engineering (Bangladesh) undertaking a project funded by the UK Department for International Development (DFID). Any views expressed are not necessarily of DFID (Permit Number IPR/88-20CGC)

Contents

CHAPTER 1: INTRODUCTION 1
   Historical perspectives 1
   We are what we eat and drink 5
   Deficiencies, excesses and imbalances of trace elements 8

CHAPTER 2: GEOCHEMISTRY OF THE TROPICAL
                      ENVIRONMENT 19
   Tropical environment 19
      Arid zone 20
      Seasonally dry tropics and sub-tropics 21
      Humid tropics and sub-tropics 21
      Mountainous zone 22
   Rock weathering and soil formation in the tropics 24
      Tropical weathering of mineralized terrains 29
      Weathering profiles 30
      Weathering of nickeliferous serpentinites 30
      Formation of secondary minerals 32
      Chemistry of weathering of ultra-basic rocks 33
   Hydrogeochemistry of the tropical environment 35

CHAPTER 3: BIOAVAILABILITY OF TRACE ELEMENTS
                      AND RISK ASSESSMENT 47
   Bioaccumulation 47
   Bioavailability 48
   Risk assessment 51
   Aspects of epidemiology in medical geology 54
   Causation and correlation 55
   Homeostasis in medical geology 56

CHAPTER 4: MEDICAL GEOLOGY OF FLUORIDE 59
   Geochemistry of fluoride 60
      Geochemistry of fluoride in weathering and solution 62
      Fluoride in soils 66
      Fluoride in sediments 66
      Fluoride in plants 67

Fluorides and health 69
　Bioavailability of fluoride 69
　Dental fluorosis 71
　Skeletal fluorosis 76
Case studies 78
　Dental fluorosis in Sri Lanka 78
　　Distribution of fluoride in the groundwater of Sri Lanka 81
　Dental fluorosis in India 84
　Fluorosis in the east African rift valley 87
　Endemic fluorosis in China 92
　　Brick tea fluorosis in China 94
Defluoridation of high fluoride groundwater 95

CHAPTER 5: IODINE GEOCHEMISTRY AND HEALTH 99
The iodine cycle in the tropical environment 99
　Iodine sorption on clays and humic substances 109
　Effect of microbial activity on iodine geochemistry 111
　Iodine in drinking water 112
　Iodine in food 112
　Plate tectonics, high altitudes and iodine cycling 114
Iodine and health 117
Iodine Deficiency Disorders (IDD) 117
　　Endemic cretinism 120
　　Goitrogens 121
　Endemic goitre in Sri Lanka 125
　The Endemic goitre belt of India and Maldives 130
　Goitre in Vietnam 132
　Iodine deficiency in China 132
　Iodine deficiency in East Africa 135

CHAPTER 6: NITRATES IN THE GEOCHEMICAL
　　　　　　　　ENVIRONMENT 139
The nitrogen cycle 139
Nitrates, fertilizers and environment 142
Nitrogen loading in rice fields 147
Nitrates from human and animal wastes 148
Nitrates and health 153
　Nitrates and methaemoglobinaemia 153
　Nitrates and cancer 154

CHAPTER 7: MEDICAL GEOLOGY OF ARSENIC 157
  Introduction 157
    Arsenic in rocks and minerals 161
    Arsenic in soils 161
    Arsenic in natural waters 164
    Arsenic adsorption and desorption 168
    Microorganisms and their impact on arsenic speciation
      and mobility 169
  Medical geology of arsenic- the West Bengal,
    Bangladesh example 175
  Bangladesh basin-geography and geology 175
    Sediment characteristics 176
    Mineralogy and geochemistry of sediments 178
    Organic matter 179
    The scale of the problem 179
  The geochemical mechanisms of arsenic mobility in the
    Bengal basin 181
    Distribution of arsenic in the aquifer system 181
    Geochemical mechanism of arsenic mobility 183
    Arsenic in rice and other crops 185
  Health effects of arsenic 186

CHAPTER 8: WATER HARDNESS IN RELATION TO
               CARDIOVASCULAR DISEASES AND
               URINARY STONES 191
  Water hardness 192
    Cardio-protective role of calcium and magnesium 192
  Geochemical basis for tropical endomyocardial fibrosis (EMF) 197
  Effect of water hardness on urinary stone formation (urolithiasis) 200
    Types of stones 202
      Calcium oxalate 202
      Calcium phosphate 202
      Uric acid 202
      Magnesium ammonium phosphate stones 202
      Cysteine 203

CHAPTER 9: SELENIUM- A NEW ENTRANT TO
               MEDICAL GEOLOGY 205
  The geochemistry of selenium in the environment 205
    Microbial transformation of selenium 211
    Dissimilatory reduction 212
    Assimilatory reduction 213
    Oxidation 215
    Methylation and volatilization 215

Selenium and human and animal health 216
    Immune function 217
    Viral infection- AIDS 217
    Reproduction 217
    Mood 217
    Thyroid function 218
    Cardiovascular diseases 218
    Oxidative-stress or inflammatory conditions 218
    Cancer 218
Selenium deficiency diseases in China 219
Selenium and iodine deficiency diseases (IDD) 222

CHAPTER 10: GEOLOGICAL BASIS OF PODOCONIOSIS, GEOPHAGY AND OTHER DISEASES 223
Geophagy 223
    Geophagy among animals 227
Ingestion of geomaterials for human health-the medical concerns 229
Podoconosis-a geochemical disease 231
Natural dust and pneumoconiosis 234

CHAPTER 11: HIGH NATURAL RADIOACTIVITY IN SOME TROPICAL LANDS – BOON OR BANE? 237
Terrestrial radiation in beach sands in Brazil 238
Monazite rich beach sands of India 240
High natural radioactivity of the Minjingu phosphate mine,
    Tanzania 243
Very high natural radiation in Ramsar, Iran 243
High natural background radiation in Yangjiang, China 245
The Oklo natural reactor 245
Radiation and health 247

CHAPTER 12: BASELINE GEOCHEMICAL DATA FOR MEDICAL GEOLOGY IN TROPICAL ENVIRONMENTS 251
Geochemical mapping - China's example 252
Soil micronutrient maps in tropical countries and medical
    geology 255
Future prospects for medical geology 256

References 259
Index 293

About the Book

Over two billion people live in tropical lands. Most of them live in intimate contact with the immediate geological environment, obtaining their food and water directly from it. The unique geochemistry of these tropical environments has a marked influence on their health, giving rise to diseases that affect millions of people. The origin of these diseases is geologic as exemplified by dental and skeletal fluorosis, iodine deficiency disorders, trace element imbalances to name a few. This book, one of the first of its kind, serves as an excellent introduction to the emerging discipline of Medical Geology.

About Authors

Professor C. B. Dissanayake

Chandrasekara Bandara Dissanayake received his B.Sc. degree with a first class from the University of Ceylon in 1970 and his D.Phil degree in geochemistry from Oxford University, UK in 1973. He then returned to Sri Lanka and worked at the University of Peradeniya. In 1991, he was awarded the D.Sc degree by Oxford University for his internationally acclaimed work in Applied Geochemistry. At present he is a Senior Professor of Geology at the University of Peradeniya and the Director, Institute of Fundamental Studies, Sri Lanka. Prof. Dissanayake is a Fellow of the Third World Academy of Sciences and the National Academy of Sciences of Sri Lanka. In 1999 he was honoured by the President of Sri Lanka with the highest national title for science for his outstanding contributions to the Geology of Sri Lanka. Further, he has received the National Award for Scientific Achievement and the Gold Medal of the Institute of Chemistry, Ceylon. He is a recipient of the Fellowship of the Alexander von Humboldt Foundation, Germany and has been a Visiting Professor in Germany and France. Among his research interests are medical geology, hydrogeochemistry, geochemistry of mineral deposits in high-grade metamorphic terrains, environmental geology and Gondwana geology. He is the author of over 200 research publications and several books.

Dr. Rohana Chandrajith

Rohana Chandrajith received his Bachelor of Science Degree in Geology from the University of Peradeniya, Sri Lanka in 1988 and his Master of Science Degree from Shimane University, Japan. Supported by the German Academic Exchange Services (DAAD), he carried out his doctoral research on exploration geochemistry from 1996 to 1999 and received his Ph.D. degree from the University of Erlangen-Nürnberg with an excellent pass. He then worked at the Sabaragamuwa University of Sri Lanka from 1999 to 2002 and subsequently joined the University of Peradeniya, as a Senior Lecturer. In 2006, he received a Georg Forster Research Fellowship from the Alexander von Humboldt Foundation of Germany and worked at the University of Erlangen-Nürnberg in the field of Medical Geology. Dr. Chandrajith's research interests cover a wide range of topics in geosciences particularly on medical geology and aqueous geochemistry. His research work has resulted in over 45 research papers in international journals and two books.

CHAPTER 1

INTRODUCTION

HISTORICAL PERSPECTIVES

At the launch of the journal "Chemical Geology" in 1966, Manten (1966) traced the historical foundations of chemical geology and geochemistry. The first use of the word "geochemistry" was attributed to the Swiss chemist C.F. Schönbein in 1838. He had discovered ozone and worked as a Professor at the University of Basel. Geochemistry at this period of time emphasized the need to investigate the chemical and physical properties of geological materials and their age relationships.

W. J. Vernadsky, a Russian geochemist developed the subject of geochemistry further by studying the concentration of chemical elements in relation to the crust of the earth and the earth as a whole. Vernadsky's contributions to geochemistry are well acknowledged and it is mainly due to his early work that geochemistry in Russia received international recognition. His influence on the younger colleagues was also marked and research on geochemistry intensified with the publication of a large number of research papers.

V.I. Verndasky, another Russian scientist, made major contributions to geochemistry and published the book "La Geochemie" in 1924. Other Russian scientists such as A.E. Fersman and A.P. Vinogradov developed the science of geochemistry still further and their research findings opened up whole new disciplines related to geochemistry, such as space chemistry, biogeochemistry and chemistry of the interior of the earth.

Modern Geochemistry had its foundations in the outstanding work of F.W. Clarke and V.M.Goldschmidt. Clarke's studies on rocks as chemical systems, the nature of chemical changes caused by external agencies and the establishment of equilibrium were indeed most significant. His publication

"The Data of Geochemistry" which appeared in several editions from 1908 to 1924 was a landmark contribution.

V.M. Goldschmidt, referred to as the father of modern geochemistry by many, classified the elements geochemically and laid the theoretical foundations for the understanding of the affinities of different chemical elements to the different units of the earth. He established the rules governing the distribution of the elements in minerals and rocks. He studied the geochemistry of the rare-earth elements (REE) and recognized the lanthanide contraction. The importance of ionic size, ionic charge and the nature of chemical bonding in minerals was well recognized by Goldschmidt and his understanding of the geochemical cycles and pathways was truly remarkable.

Moreover, Goldschmidt's studies on meteorites and their importance in the partitioning of elements in the crust, mantle and core of the earth led to further studies of the chemistry of meteorites. It is now recognized that plate tectonics and planetology with their far reaching transformation of geological concepts had their roots in the basic discoveries of physics and chemistry made between 1900 and 1930 (Wedepohl, 1996). A special publication of the Geochemical Society, paid tribute to Viktor Moritz Goldschmidt-the truly outstanding geochemist of the century (Mason, 1992). His book "Geochemistry" (Goldschmidt, 1954) remains a classic work in the field of geochemistry.

Subsequent workers of the modern era such as N.L. Bowen, S.S. Goldich, K. Rankama and T. Sahama, H.L. Barnes, H.C. Helgeson, H.D. Holland, K. Krauskopf, B. Mason, W.S. Fyfe, K.H. Wedepohl, I.M. Garrels and C. Christ among many others, made invaluable contributions to the different aspects of the major discipline of geochemistry. With the availability of extremely powerful and accurate analytical tools, the science of geochemistry is now probing into the chemical compositions of materials in planets as far as Mars and Jupiter as well as microbial matter from the interior of the earth. With the development of the science of geochemistry, our understanding of the various earth processes has improved vastly. The knowledge of the geochemical cycles and their pathways transcending the different spheres namely atmosphere, hydrosphere, biosphere and lithosphere enabled scientists to discover the intricate processes which influence the distribution of the chemical elements. The significance of the interactions between the different spheres is of special interest to geoscientists. From among these, the geosphere-biosphere interactions have aroused the curiosity of many and the role of geology on biological processes is beginning to be increasingly investigated.

With the establishment of a relationship leading to rock-soil-water-plant-animal interactions, the influence of geology on animal and human health soon became apparent. As far back as the 11th Century AD, the Chinese recognized this influence. Regional variations in the prevalence of human diseases were observed, indicating a geological association with the incidence of some of these diseases. The science of Medical Geology was thus established. The importance of the influence of the composition of the geological materials such as rocks, minerals, soils and water on the aetiology of a localised disease is better appreciated when man and animal are considered as only a part of a system in a total environment. The geochemical cycles and element pathways also involve man and hence the influence of the chemical composition on human and animal health becomes clearer.

Terms such as Geomedicine, Medical Geology, and Medical Geochemistry with their subtle differences in definitions, broadly deal with the geographical distributions of diseases with geological materials playing a very significant part. As in many other scientific disciplines, the birth of a new discipline is a gradual process with early examples arising from ancient chronicles, reports and hear-say accounts.

Diseases such as the iodine deficiency disorders (IDD) and their relationship to the geochemistry and geographical distribution of iodine in soil and water had been recognized as far back as 1851 by Chatin (1852), a French Chemist. Likewise, several other geographically related diseases, as exemplified by dental and skeletal fluorosis were shown to be closely related to the geological environment.

The term "Medical Geochemistry", though not as commonly used as Medical Geology, could perhaps be broadly described as the science dealing with the chemistry of the elements in geological materials in relation to health in man and animals. An understanding of the chemical species, their pathways, toxicities and bioavailabilities in human and animal health and their impact comprise the main subject matter of the discipline of Medical Geochemistry. Unlike the much broader discipline of Medical Geology, the subject of Medical Geochemistry emphasizes the geochemistry of the chemical species concerned and their pathways. It also overlaps with other fields such as biochemistry and molecular biology. The mechanisms of the entry of the trace elements from the geosphere into the human and animal tissues is of particular scientific interest and has aroused the curiosity of scientists in a variety of disciplines. Medical geochemistry therefore deals essentially with the geochemistry of those trace elements which originate in the geological environment and which has a profound impact on the human and animal life living in close proximity. It is thus closely related to the geographical distribution of the diseases concerned, in view of the fact that the geology and hence the geochemistry of a terrain concerned is unique to that terrain. The excesses and deficiencies of the trace elements in the rocks and minerals of the terrain are shown by their geochemistry and a thorough understanding of the chemical speciation and the geochemical cycles is clearly warranted.

The field of medical geochemistry has benefited immensely from the pioneering work of Russian, Scandinavian and the British geochemists. The Russian geochemists (Vinogradov, 1938; Kovalsky, 1979) in particular had recognized "biogeochemical provinces" in the former USSR where they established relationships between geochemistry and health in both humans and animals. Also of importance is the early work of Låg (1980, 1983), in the general area of "geomedicine" which was explored fully in two symposia under the Norwegian Academy of Science and Letters.

In Canada, Warren et al. (1972) made important contributions to the field of medical geology, where they recognized the importance of the geological distribution of trace elements and their impact on health. British geochemists also made major contributions to this field as evidenced by the work of the Royal Society of London (1983) and Bowie and Thornton (1985). The publication of the National Academy of Sciences (1974, 1978) on "geochemistry and the environment" was a result of the workshops undertaken by its sub-committee on geochemical environment in relation to health and disease. More recently, The British Geological Survey has made extremely valuable contributions to the field of Medical Geology through their detailed studies in different terrains of the world. The work of Jane Plant and her team deserves special mention.

Recently, however, with the development of fast and accurate analytical techniques such as X-ray fluorescence (XRF) and Inductively Coupled Plasma-Mass spectroscopy (ICP-MS) and also with concurrent improvements in computer technology, geochemists are now able to geochemically map vast areas fairly rapidly and superimpose geographic information data. This has led to the better understanding of the geochemical pathways through the various spheres, notably the biosphere. The science of "Medical Geology" can now be considered as a well established discipline in its own right.

WE ARE WHAT WE EAT AND DRINK

From a purely geochemical point of view, this old adage has a scientific basis. Even though there are complicating factors; such as life style, sex, age, migrations and food habits; deficiencies, excesses or imbalances in the supply of inorganic elements do exert a marked influence on human and animal health and also on the susceptibility to disease. As shown by Mills (1996), these in turn, are frequently attributable to the composition of the geochemical environment as modified by the influence of soil composition and botanical or cultural variables upon the inorganic composition of the diet. Trace element anomalies have a significant impact on the composition of food chains.

Figure 1.1 illustrates the geochemical pathways which enable the trace elements to enter the human body. The general geochemical principles govern the processes of element distribution, accumulation and depletion of trace elements in specific environments and this leads to situations where humans and animals encounter mineral excesses, deficiencies or imbalances. Such scenarios are somewhat clearer in developing countries where large populations live in close contact with the soil and immediate environment and whose food chains and trace element inputs are heavily dependent on the geochemistry of the habitat. Marked geochemical variations and anomalies therefore clearly influence the health of the population.

The importance of medical geology on the application to human and animal health can be emphasized by the following examples as summed up by Dissanayake and Chandrajith (1999).

i. More than 30 million people in China alone suffer from dental fluorosis caused by the excess of fluoride in drinking water. This is clearly related to the geochemistry of the groundwater among some other factors. Many countries, such as South India, Sri Lanka, Ghana, Tanzania among others also have very high incidences of dental (and in some cases) skeletal fluorosis.

ii. Nearly 1 billion people (notably in developing countries) suffer from Iodine Deficiency Disorders (IDD) caused by the lack of iodine in the diet. These diseases include endemic goitre, cretinism, foetal abnormalities among others. The relationship between the geochemistry of iodine in the rocks, soil, water, sea and atmosphere on the incidence of IDD is one of the most interesting research studies that is now creating global interest among scientists.

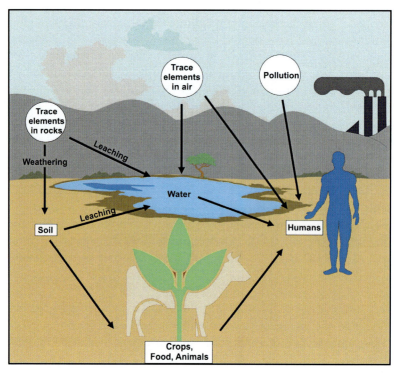

Fig. 1.1. Geochemical pathways of trace elements entering the human body

iii. Arsenic is a toxic and carcinogenic element present in many rock-forming minerals including iron oxides, clays and in particular sulphide minerals. When this arsenic gets into the groundwater through oxidation and subsequently into the human body through drinking water, serious health hazards can occur. Well documented cases of chronic arsenic poisoning are known in southern Bangladesh, West Bengal (India), Vietnam, China, Taiwan, Chile, Argentina, and Mexico. Skin diseases are the most typical symptoms of chronic exposure to arsenic in drinking water, including pigmentation disorders; hyperkeratosis and skin cancer, but other renal, gastrointestinal, neurological, haematological, cardiovascular and respiratory symptoms can also result. The study of the medical geochemistry of arsenic is now being recognised by several governments as a priority area of study.

iv. Recent evidence indicates that cancer, after heart diseases, is the leading killer in many industrialized societies and is largely due to environmental factors. A large number of causative factors which have been isolated are in one way or other environmental. Historically research into the causes of cancer was often based on the hypothesis that all cancers are environmentally caused until the contrary was proved. Geochemistry therefore plays an important role in the aetiology of cancer. A good example from developing countries that affects millions of poor people is the contamination of drinking water by nitrogenous matter such as human and animal wastes, nitrogen containing fertilizers etc. The common diseases caused by this are stomach and oesophageal cancer and methaemoglobinaemia ('blue baby' syndrome), caused by excess nitrates. The passage of these chemical species from the environment into the food chain and into the human body is mostly geochemical and the medical geochemistry of cancer has developed into an intriguing field of research (Dissanayake and Weerasooriya, 1987).

v. Podoconiosis or non-filarial elephantitis, named and characterised by Price (1988) affects large populations in Ethiopia, Kenya, Tanzania, Rwanda, Burundi, Cameroon and the Cape Verde Islands. The most interesting feature observed was that the affected areas were consistently associated with red clay soils. Analysis of lymph nodes from diseased tissues showed the presence of microparticles consisting predominantly of aluminium, silicon, and titanium. It was suggested that the pathological agent is a mineral from volcanic bedrocks, probably the amphibole eckermanite (Harvey et al., 1996). In this case too, the importance of research into medical geology is obvious.

vi. The main causes of low production rates among grazing livestock in many developing countries are probably linked to under nutrition. However, mineral deficiencies and imbalances in forages also have a negative effect. The assessment of areas with trace element deficiency or toxicity problems in grazing livestock has traditionally been executed by mapping spatial variations in soil, forage, animal tissue or fluid compositions. Regional stream sediment geochemical data sets collected principally for mineral exploration already exist in many developing countries (Agget et al., 1980; Plant and Thornton, 1986). The application of these data sets for animal health studies in tropical regimes is now being developed (Fordyce et al., 1996).

vii. One of the most intriguing yet, not very well defined aspects is the geochemical correlation between the incidence of cardiovascular diseases and the water hardness in the areas concerned. In several countries and areas, a negative correlation has been observed between water hardness of the country or region and its death rate due to heart diseases (Masironi, 1979). Even though a causal effect still cannot be ascribed to this geochemical correlation, the effect of trace elements in drinking water on heart diseases is worthy of serious study. It is of interest to note that such a negative association between water hardness and cardiovascular pathology is evident in both industrialized and developing nations.

DEFICIENCIES, EXCESSES AND IMBALANCES OF TRACE ELEMENTS

Geochemically, elements are often classified as major, minor and trace elements depending on their relative abundance in geological materials, even though a precise definition has not been ascribed. Elements are broadly classified as essential or toxic depending on their impact on human and animal health. With new research and discoveries, the optimum and danger levels often change as exemplified by mercury and arsenic when their extreme health hazards were realized. As against this, with advances in analytical techniques as well as our understanding of the physiological importance of the different elements, new elements hitherto thought of as not being physiological important, have recently been classified as essential.

Such a classification of elements essential to human health and those considered to be toxic or undesirable is shown in Figure 1.2. What is of great importance is the fact that, whatever the element, it is the dosage that is critical. Elements, which are considered as being truly beneficial to human and animal health, may also lead to debilitating diseases if ingested in large doses.

Introduction

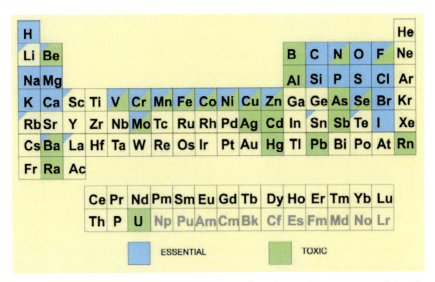

Fig. 1.2. The periodic table of elements showing those elements essential to human health and those considered or known to be toxic or undesirable. Note that some elements fall in both categories, others are possibly essential for living organisms (Source: Groundwater Geochemistry and Health, 1996; reproduced with kind permission of the British Geological Survey)

It is of interest to note that Paracelsus (1493–1541) defined the basic law of toxicology, namely *"All substances are poisons; there is no element which is not a poison. The right dose differentiates a poison and a remedy"*. This basic law can be put in its perspective as shown in Figure 1.3. The effects of deficiency and excess of essential and non-essential trace elements on the growth and health of organisms are highlighted. It should be noted that it is the optimal range of the essential elements that one needs to keep intact in the ideal situation. Diseases such as hyperkalemia (excess K), hypercalcaemia (excess Ca) and hyper-phosphatemia (excess P) are known.

Potassium for example, is a major ion of the body, 98% of which is intracellular. The concentration gradient is maintained by the sodium and potassium-activated adenosine triphosphatase (Na^+/K^+-ATP ase) pump. The ratio of intracellular to extra-cellular potassium is important in determining the cellular membrane potential. Small changes in the extra-cellular potassium level can have profound effects on the function of the cardiovascular and neuromuscular systems. It has been shown that in serum the normal potassium level is 3.5-5.0 meq/L and the total body potassium stores are approximately 50 meq/kg (3500 meq in a 70 kg person). Hyperkalemia is therefore defined as a serum potassium level greater than 5.5 meq/L, bearing in mind that a level of 7.0 meq/L and greater is classified as a severe condition, with a mortality rate as high as 67% if not treated rapidly (Garth, 2001).

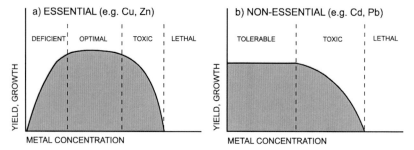

Fig. 1.3. Deficiency and oversupply of essential and non-essential trace elements (Förstner and Wittman, 1981)

Similarly calcium, another essential element, could cause the disease hypercalcemia (Carrol and Schode, 2003). The reference range of serum calcium levels is 8.7-10.4 mg/dl, with somewhat higher levels present in children. Approximately 40% of the calcium is bound to protein, primarily albumin, while 50% is ionized and is in physiologically active form. The remaining 10% is complexed to anions.

For hypercalcemia to develop, the normal calcium regulation system must be overwhelmed by an excess of parathyroid hormone (PTH), calcitriol, some other serum factor that can mimic these hormones or a huge calcium load (Fukagawa and Kurokawa, 2002).

Iron, another essential and important nutrient, can cause significant damage to the endothelium, the inner lining of blood vessels, if present in excessive levels in the body. Table 1.1 shows examples of effects of mineral deficiencies and excesses in human beings and Table 1.2, examples in plants.

While food is a major source of the trace elements needed by the body, some other elements, notably fluorine, is mainly ingested through water as fluoride. Figure 1.4 illustrates the levels of trace elements, both essential and toxic, in groundwater. Their significance in terms of health and environmental protection is of particular importance in that provision of safe drinking water is a dire need in developing countries.

Introduction

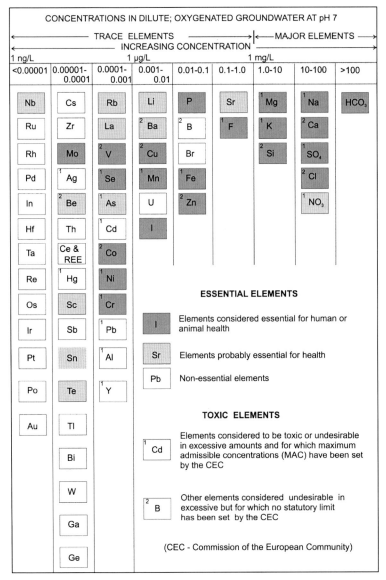

Fig. 1.4. Major and trace elements in groundwater and their significance in terms of health. Concentrations shown are those typical of dilute oxygenated groundwater at pH 7 (Edmunds and Smedley, 1996).

The World Health organization (WHO) plays a leading role in providing guidelines for food, drinking water quality and also for trace elements. Table 1.3 shows the guidelines for the different elements and other harmful substances in drinking water. According to the WHO (1996), the traditional criteria for essentiality of trace elements for human health are that absence or deficiency of the element from the diet produces either functional or structural abnormalities and that the abnormalities are related to, or a consequence of, specific biochemical changes that can be reversed by the presence of the essential metal.

As pointed out by the WHO, one of the important aspects of the essentiality of trace elements that the medical geologist must necessarily be aware of, is the margin between individual and population requirements and the tolerable intake (TI). This may turn out to be very small and in some instances may even overlap among individuals and populations. Developing countries with their different life styles and inadequate nutritional requirements coupled with a host of natural and anthropogenically related environmental problems may have markedly different tolerable intakes as compared to those living in developed and temperate countries. Medical geology should therefore necessary link up with toxicology and nutrition science. Terms such as acceptable daily intake (ADI), tolerable intake (TI), tolerable upper intake level (UL) and homeostasis as used by the International Programme on Chemical Safety (IPCS, 1987) need to be incorporated into the terminology of Medical Geology. Table 1.4 shows the Indian water standards which perhaps may be more appropriate for those living in developing countries of the humid tropical belt.

Groundwater quality, drinking water, health and sanitation are of special importance to developing countries in view of the fact that central water purification and proper water disposal techniques may not be available in most areas and the groundwater quality then becomes critically important in locating sites for wells for domestic and community use. In highly populated countries such as India and Sri Lanka lying within the tropical belt, water quality becomes a major factor in urban and sub-urban development.

An interesting case study in this regard is highlighted by Hutton and Lewis (1980) in their study of nitrate pollution of groundwater in Botswana. They observed nitrate levels as high as 603 mg/L in several water supplies providing drinking water to many villages. A lithium chloride tracer injected into a pit latrine was detected in the supply borehole 25 m away after only 235 minutes. The steep hydraulic gradient between the latrine and the borehole had obviously induced the rapid movement of nitrates in open fissures.

Table 1.1. Typical geochemical and soil features associated with inorganic element anomalies causing nutritional diseases in man and domesticated livestock (Mills, 1996)

SYNDROME	ENVIRONMENTAL QUALITY	SPECIES AFFECTED
Article I **Deficiencies**		
low cobalt	Soils intrinsically low in Co, eg. extensively leached, acid arenaceous soils or with Co immobilized with Fe/Mn hydroxide complexes.	ruminant livestock, specific
low phosphorus	High Fe/Al parent materials with low pH and highly organic soils.	ruminant livestock, specific
low selenium	Soils intrinsically low in Se, eg: leached arenaceous soils particularly when low in organic and argillaceous fractions.	human subjects, farm livestock, general
low zinc	Calcareous parent materials and derived soils especially when adventitious soil present in diets high in cereals and legumes. Arid arenaceous soils.	human subjects farm livestock, general
Article II **Toxicities**		
high arsenic	Waters from hydrothermal sources or soils derived from detritus of mineral ore (especially Au) workings. Well waters or irrigation waters from sandstones high in arsenopyrite.	
high fluoride	Waters from some aquifers especially from rhyolite-rich rocks, black shales or coals; soils from F-containing residues of mineral or industrial deposits. Aggravated by high evaporative losses.	human subjects farm livestock, general
high molybdenum	Mo from molybdeniferous shales or local mineralization especially if drainage is poor and soil pH >6.5 (a significant cause of secondary Cu deficiency).	ruminant livestock, specific
high selenium	Bioaccumulation of Se in organic-rich soil horizons; accumulation by high evaporative losses of high pH groundwaters.	human subjects farm livestock, general

Table 1.2. Signs of deficiency or excess of mineral elements in plants (Wellenstein and Wellenstein, 2000)

ELEMENT	PRIMARY FUNCTIONS IN PLANT	SIGNS OF DEFICIENCY	SIGNS OF EXCESS
Nitrogen	Growth of green (leaf and stem) portions of plant.	Reduced growth, vigour. Chlorosis of older leaves first, premature leaf drop.	Soft growth, spindly growth, leaf curl, reduced flowering, symptoms of Potassium deficiency.
Potassium	Root growth, sugar and starch production, cell membrane integrity.	Dwarfing, chlorosis of older leaves first, leaf curling.	Deficiency symptoms of nitrogen, magnesium, calcium, iron, zinc, copper, manganese.
Calcium	Cell wall formation, cell division, enzyme catalyst, neutralization of toxic metabolites.	Poor growth, deformed newer leaves, chlorosis of newer leaves, blackened areas at leaf ends and new growths with a leading yellow edge, stunted, shortened roots, dead root tips.	Symptoms of magnesium deficiency.
Magnesium	Chlorophyll and protein production, carbohydrate metabolism, enzyme activation.	Interveinal and marginal chlorosis starting in the older leaves, increase in appearance of anthocyanin in leaves, necrotic spotting.	Symptoms of calcium deficiency.
Phosphorus	Constituent of nucleic acids, coenzymes NAD and NADP required for photosynthesis, respiration and many metabolic processes, and the energy compound ATP. Essential for root growth, flowering and seed production.	Older leaves are affected first, an increase in anthocyanin pigment and a dark blue green coloration, sometimes with necrotic areas, and stunting.	Symptoms of nitrogen, zinc, iron deficiencies.
Sulfur	Protein formation, photosynthesis, and nitrogen metabolism.	Root stunting, general chlorosis starting with younger leaves.	

ELEMENT	PRIMARY FUNCTIONS IN PLANT	SIGNS OF DEFICIENCY	SIGNS OF EXCESS
Boron	Sugar transport, DNA synthesis.	Death of meristematic tissue, root stunting, no flower formation.	Interveinal leaf necrosis.
Iron	Component of cytochromes and ferrodoxin, synthesis of chlorophyll.	Interveinal chlorosis of newer leaves.	
Manganese	Enzyme activation in respiration and nitrogen metabolism.	Interveinal chlorotic and necrotic spotting.	Stunting, necrotic spotting of leaves.
Zinc	Trytophan synthesis, enzyme activation.	Smaller, distorted leaves, stunting, interveinal chlorosis of older leaves, white necrotic spotting, resetting.	Symptoms of magnesium or iron deficiency.
Copper	Enzyme component, electron carrier protein in chloroplast.	Stunted, misshapen growth.	Symptoms of magnesium or iron deficiency.
Molybdenum	Nitrogen and potassium metabolism.	Chlorotic interveinal mottling, marginal necrosis, folding of the leaf, no flower formation.	

Such case studies clearly illustrate the overlap between a variety of disciplines such as structural geology, hydrology, community health, town and country planning and medical geology.

As a tool in assessing the state of the environment, geochemical parameters in relation to human health issues have proven to be of immense value, both nationally and internationally. The availability in many parts of the world of sufficient data that show trends over time make such parameters useful indicators in monitoring environmental changes. Trace elements and chemical species that show direct relationships are the most useful as geoindicators of human health (Dissanayake, 1996). Medical geology is therefore rapidly gaining status in monitoring environmental health.

Table 1.3. Characteristics of water and waste water. WHO (1993) Drinking water guidelines (mg/L unless specified)

Parameter	Guide Level	Note
Microbiological		
Total Coliforms/100 ml	0	95% absent over 12 month period
E.coli	0	
Inorganics		
Antimony	0.005	
Arsenic	0.01	
Barium	0.7	
Boron	0.3	
Cadmium	0.003	
Chromium	0.05	
Copper	2	
Cyanide	0.07	
Fluoride	1.5	Depends on local conditions
Lead	0.01	
Manganese	0.5	
Mercury	0.001	
Molybdenum	0.07	
Nickel	0.02	
Nitrate	50	Sum of ratio of concentration to GL
Nitrite	3	should not exceed 1 for both together
Selenium	0.01	
Organics (Part list only)		
Carbon tetrachloride	0.002	
Dichloromethane	0.020	
Trichloroethene	0.040	
Benzene	0.010	
Toluene	0.700	
Ethylbenzene	0.300	
Acrylamide	0.0005	
Nitrilotriacetic acid	0.200	
Tributyltin oxide	0.002	
Pesticides (Part list only)		
Atrazine	0.002	
DDT	0.002	
2,4-D	0.030	
Heptachlor	0.00003	
Pentachlorophenol	0.009	
Permethrin	0.020	
Simazine	0.002	
Mecoprop	0.010	

Table 1.4. Indian water standards (Bureau of Indian Standards, 1991)

Property/ Constituent	Desirable limit	Permissible limit	Undesirable effect outside the desirable limit
Physio-chemical Characteristics			
Turbidity JTU Scale	2.5	10	Aesthetically undesirable
Colour (Pt-Cobalt Scale)	5.0	25	Aesthetically undesirable
Taste and Odour	Unobjectionable	Unobjectionable	Aesthetically undesirable
Major Chemical Constituents			
pH	6.5-8.5	6.5-9.2	Affects taste
Total dissolve solids, mg/L	500	1500	Causes gastrointestinal irritation.
Total Hardness, as $CaCO_3$ mg/L	300	600	May cause urinary concretion, disease of kidney, bladder and stomach disorder.
Calcium, mg/L	75	200	Essential for nervous and muscular system, cardiac function and coagulation of blood. Deficiency causes rickets. Excess concentration causes kidney or bladder stone and irritation in urinary passage.
Magnesium, mg/L	<30 if SO_4 is 250 mg/L	100	Essential as an activator for many enzyme systems. Deficiency results in structural and functional changes. Excess concentration may have laxative effects. Magnesium salts are cathartic and diuretic.
Chloride, mg/L	250	1000	Affects taste and palatability. Causes indigestion may be injurious to people suffering from heart and kidney diseases.
Sulphate, mg/L	200	400	Causes laxative effects in presence of Magnesium.
Nitrate, mg/L	45	100	Causes infant Methemoglobinemia (Blue babies). May cause gastric cancer and affects central nervous system and Cardio-Vascular system.
Fluoride, mg/L	1.0	1.5	Essential for teeth and

Property/ Constituent	Desirable limit	Permissible limit	Undesirable effect outside the desirable limit
			bones, reduces dental caries in concentration range of 0.8 - 1.0 mg/L and at high level teeth mottling, skeletal and crippling fluorosis occurs.
Iron, mg/L	0.3	1.0	Gives bitter sweet astringent taste.
Manganese, mg/L	0.05	0.5	Unpleasant taste
Copper mg/L	0.05	1.5	Astringent taste, deficiency results in nutritional anaemia in infants, high concentration may damage liver and cause central nervous system irritation and depression.
Zinc, mg/L	5.0	15	Very small amount beneficial. Imparts astringent taste at higher concentration.
Toxic Constituents			
Arsenic, mg/L	0.05	0.05	Skin diseases, circulatory system problem, risk of cancer.
Cadmium, mg/L	0.01	0.01	Kidney damage.
Chromium(VI), mg/L	0.05	0.10	Lung tumor, Allergic dermatitis
Cyanide, mg/L	0.05	0.05	Causes nerve damages and thyroid problem.
Lead, mg/L	0.05	0.05	Serious Cumulative body poison
Selenium, mg/L	0.01	0.10	Small amount beneficial, large amount toxic.
Mercury, mg/L	0.001	0.001	Large amount causes brain and kidney damage
Polynuclear Aromatic hydrocarbon, mg/L	0.20	0.20	Toxic

CHAPTER 2

GEOCHEMISTRY OF THE TROPICAL ENVIRONMENT

TROPICAL ENVIRONMENT

Tropical regions cover approximately 40% of the surface of the earth and have a diversity of climates. As noted by Köppen (1936), two broad categories of tropical climates can be distinguished, namely tropical rainforest and periodically dry savannah. Depending on the rainfall, these two categories are often further subdivided.

The Food and Agricultural Organization (FAO, 1978) proposed a more detailed climatic characterization defining agro-ecological zones for the developing world. Subsequently, the FAO (1993) simplified and broadened the agro-ecological zoning method to illustrate the distribution of the major soil resources globally (Fig. 2.1). In the context of the tropical environment, only four subdivisions were considered relevant:

(a) Arid zone.
(b) Seasonally dry tropics and sub-tropics.
(c) Humid tropics and sub-tropics, and
(d) Mountainous zone.

From the point of view of medical geology, an agriculture-based classification becomes useful on account of the rock-soil-water-plant-human relationships.

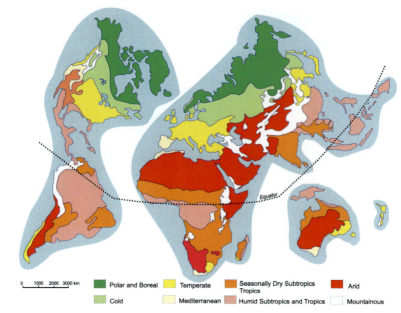

Fig. 2.1. World's major climates (FAO, 1978)

Subdivisions of the Tropical Environment

Arid zone

The arid climatic zone is characterized by an length of growing periods (LGP) of less than 75 days. The major arid areas include the Sahara, the Kalahari, the Namibian desert and the Horn of Africa. Other areas include the Arabian Peninsula, Central Asia, the western United States, the North Western parts of Brazil, South Western part of Latin America and Central Australia.

Due to the very short LGP, agricultural crops cannot complete their normal cycle and yields, if any, are insignificant. Further, the non-availability of soil moisture for most of the year, results in soil characterisation being affected. Weathering is therefore extremely slow. From among the pedogenetic processes, migration and accumulation of soluble salts, calcium carbonate and gypsum are perhaps the only functional processes and these result in Solonchaks and Solonetz, Calcisols and Gypsisols. The extremely strong wind action causes the formation of sandy soils (Arenosols) which have a lack of finer particles. Soil horizon development is virtually non-existent under these climatic conditions.

Seasonally dry tropics and sub-tropics

This type of climate is observed in the tropics and sub-tropics where the dry season prevails between 90 to 285 days and the rainfall is not confined to the winter. As shown in Fig 2.1, this zone covers approximately 2475 million ha and includes large areas of South and Southeast Asia, Northern Australia, the major part of Africa, South and Central America between the deserts and the tropical rain forests.

As observed by the FAO (1993), the climate dependent agriculture potential of this zone is heavily determined by the length and intensity of the dry season. The relatively high temperatures prevailing in the zone and the pronounced wet season favours rock weathering and soil formation with the accumulation of silica and alumina resulting in clay formation, mostly kaolinite. As shown in Table 2.1 and 2.2, Ferralsols, Acrisols and Lixisols are more abundant in this zone. When the parent rocks are abundant in Ca and Mg, smectites form and this gives rise to Vertisols.

Humid tropics and sub-tropics

This climatic zone which covers approximately 1925 millions ha, is characterized by high temperatures throughout the year and alternating wet and dry seasons. Areas such as central and coastal West Africa, the Amazon Basin, Southeast Asia and the islands of the Pacific Ocean belong to this climatic zone. The combined hot and wet climatic features of the zone provide ideal conditions for the growth of tropical rain forests and also agricultural productivity for most of the year.

Weathering is intense and the soil profile often reaches depths of as much as 150 m. Kaolinitic clays are very common and on account of the intense rainfall, nutrients are easily leached out. Among the soils present are Ferralsols (extremely weathered soils), Acrisols (less weathered) and these cover about 57% of the land of the climatic zone (Table 2.1). Groundwater tends to be abundant and this results in the formation of Gleysols and Podsols.

Table 2.1. Extent (units of 1000 ha) of major soil groups by climatic zone in the tropics and subtropics (FAO, 1993)

	Arid	Mountainous	Seasonally dry tropics and Subtropics	Humid Tropics and Subtropics
Histosols	3410	792	12232	32449
Leptosols	419462	544330	198332	66731
Vertisols	51243	3820	222983	29012
Fluvisols	90074	4401	84360	66207
Solonchaks	140324	3608	20824	4415
Gleysols	34492	11002	111543	167704
Andosols	9418	20683	18379	20674
Arenosols	395942	6980	320140	127284
Regosols	170083	35916	52109	9391
Podzols	1366	3221	13475	11343
Plinthisols	53	255	15657	42354
Ferralsols	0	4036	231347	507217
Planosols	3762	2609	74083	6267
Solonetz	57037	5367	36771	518
Greyzems	2230	5010	0	0
Chernozems	11794	1802	0	0
Kastanozems	143513	17569	44729	459
Phaeozems	2089	8721	15249	2703
Podzoluvisols	0	17358	0	0
Gypsisols	86711	1666	54	0
Calcisols	552765	44857	47267	5430
Nitisols	2792	9996	101782	87291
Acrisols	1067	13581	238808	589386
Luvisols	165499	13541	62002	21827
Lixisols	26397	11390	366862	31697
Cambisols	503586	153299	192294	95617
TOTAL	2875109	945810	2481282	1925976

Mountainous zone

The sudden changes in altitude and varying morphology of the slopes bring about marked climatic changes. The temperature and rainfall therefore change quite significantly over short distances. There is great heterogeneity in vegetation and some of the lands are of the fragile type. From the foregoing account, it is clear that the climate is perhaps the most critical determinant of the agricultural productivity, influencing the soil formation and nutrient availability.

Table 2.2. Extent (units of 1000s ha) of main soil groups by continent (FAO, 1993)

	Europe	North & C. Asia	Australasia	North America
Leptosols	64836	710863	48789	83303
Cambisols	157288	452241	81084	169019
Acrisols	4170	148241	32482	114813
Arenosols	3806	3436	193233	25512
Calcisols	56657	95264	113905	114720
Ferralsols	0	0	0	0
Gleysols	17641	323611	421	131053
Luvisols	142658	99739	127445	179244
Regosols	26848	33169	887	283933
Podzols	213624	21825	8459	220770
Kastanozems	55598	173265	1827	153704
Lixiols	0	0	0	0
Fluvisols	40250	73327	8827	10387
Vertisols	5856	11797	90019	9120
Podzoluvisols	161684	153911	0	5470
Histosols	32824	99451	1167	934562
Chernozems	98551	90566	0	40101
Nitisols	0	0	10100	0
Solonchaks	2308	46895	16565	127
Phaeozems	9221	19948	3562	70320
Solonetz	7906	30062	38099	10748
Planosols	1877	2413	38941	6500
Andosols	4058	18361	6953	17590
Gypsisols	379	16494	0	0
Plinthosols	0	0	3895	2816
Greyzems	6090	23331	0	4462
	Africa	**South & SE Asia**	**South & C. America**	**Total**
Leptosols	381531	119408	246588	1655318
Cambisols	369232	209385	135153	1573402
Acrisols	92728	263005	341161	996600
Arenosols	462401	94530	118967	901885
Calcisols	171237	220068	24318	796169
Ferralsols	319247	0	423353	742600
Gleysols	122313	37084	86707	718830
Luvisols	17609	41332	40478	648505
Regosols	140684	64357	29093	578971
Podzols	11331	5982	5522	487513
Kastanozems	2802	0	80561	467757
Lixiols	244830	86145	105545	436520
Fluvisols	98400	57357	67687	356235

Vertisols	106126	76328	38076	337322
Podzoluvisols	0	0	0	321065
Histosols	12270	24829	9245	273248
Chernozems	0	0	0	229218
Nitisols	98510	48796	48112	205518
Solonchaks	48574	48512	24344	187325
Phaeozems	341	1855	48992	154239
Solonetz	13800	0	34652	135267
Planosols	18944	4655	56566	129896
Andosols	5270	8840	45973	107045
Gypsisols	51757	21387	0	90017
Plinthosols	7992	1711	44721	61135
Greyzems	0	0	0	33883

ROCK WEATHERING AND SOIL FORMATION IN THE TROPICS

As discussed in the earlier section, the tropical environment is characterized by seasonal heavy rainfall (exceeds 5000 mm per year in some cases), with long periods of drought and high ambient temperatures. The intensity of chemical weathering is therefore extreme and some of the tropical soils are weathered to such an extent (Fig 2.2), that potassium, an essential plant nutrient usually present in appreciable concentrations is present only in trace amounts (Fyfe et al., 1983). Coincidentally, the world's most underdeveloped countries also lie in such tropical environments where poor agricultural productivity caused by the impoverishment of soil in essential nutrients has had a marked impact on the economy of these countries over the years.

Fyfe et al. (1983) noted that soils in such regions will be dominated by minerals such as gibbsite, goethite, kaolinite and quartz. A few trace elements are expected to remain in refractory minerals such as zircon, rutile etc, but the critically important feature is that most bio-essential elements are reduced to extremely low levels. These soils were termed "impossible soils".

In another interesting study on the chemistry of some Brazilian soils, where chemical leaching of elements was extreme, Kronberg et al. (1979) observed that trace elements show a wide range of behaviour and that the major elements had severely leached. They showed that the trace element behaviour appears to be largely controlled by the dominant clay or Al_2O_3-

SiO₂ minerals or the degree of weathering. The main components (SiO₂-Al₂O₃-Fe₂O₃-H₂O) are known to be present mineralogically as quartz, kaolinite, goethite and hematite.

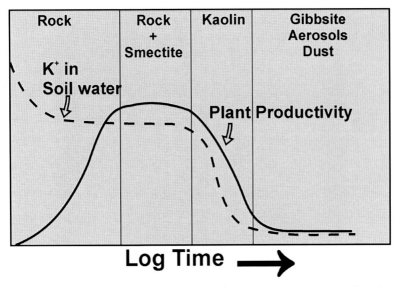

Fig. 2.2. A schematic diagram showing weathering progress from rock to bauxite and plant productivity (Fyfe et al., 1983).

Some of the tropical environments are characterized by very thick laterite profiles over 150 m thick (Fig 2.3) and it has been established that it has taken tens of millions of years for these thick laterite profiles to form. With respect to crustal abundance, except for the least soluble refractory oxides of elements such as Sc, Zr, Nb, Gd, Tm and Th and some bio-important elements such as B, Cl, Mo, Se, Sn and I, most other elements are highly depleted. Figure 2.4 illustrates the relationship of geochemistry and the kinetics of plant nutrition. Such a relationship will undoubtedly have an important bearing on the medical geochemistry of that particular environment.

The clays which form from unweathered rocks are of three types namely smectites, illites and kaolin in order of decreasing complexity. In the presence of smectite clays in association with unweathered minerals, many trace elements such as Zn, Cu, Rb, Cs, tend to get concentrated in the weathering process. According to Fyfe et al. (1983), when the primary minerals are leached and their surface areas are reduced to the point where solution concentrations drop below their metastable solubilities, the smectite and other complex clays begin to degrade with kaolin increasing in abundance in the soil. Humid tropical environments are the best locations for such processes.

Fig. 2.3. A thick laterite profile from Sri Lanka

As shown in Figure 2.4, if K_2 exceeds K_1, there will be depletion of essential elements and if K_1 ceases to operate, the levels of most essential nutrients (major and trace) will fall below the optimum value for nutrition and as a result of this infertility prevails. When the essentiality of trace elements in human and animal health are taken into consideration, an understanding of such a rock-soil-plant relationship in a terrain becomes all the more important.

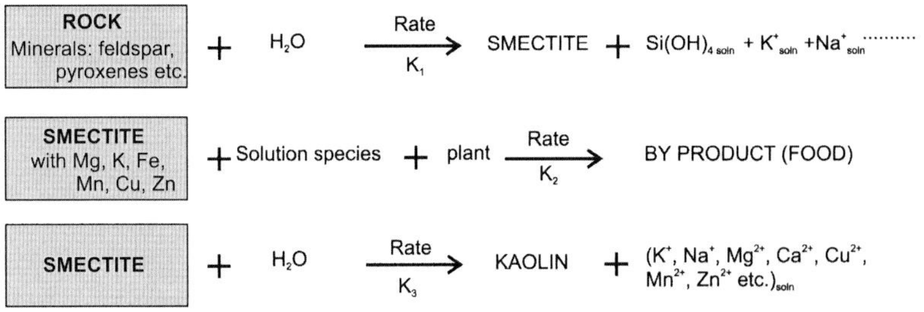

Geochemistry of the Tropical Environment

ROCK (Minerals: feldspar, pyroxenes etc.) + H_2O $\xrightarrow{\text{Rate } K_1}$ SMECTITE + $Si(OH)_{4\,soln} + K^+_{soln} + Na^+_{soln}$ ....

SMECTITE with Mg, K, Fe, Mn, Cu, Zn + Solution species + plant $\xrightarrow{\text{Rate } K_2}$ BY PRODUCT (FOOD)

SMECTITE + H_2O $\xrightarrow{\text{Rate } K_3}$ KAOLIN + $(K^+, Na^+, Mg^{2+}, Ca^{2+}, Cu^{2+}, Mn^{2+}, Zn^{2+}$ etc.$)_{soln}$

Fig. 2.4. Geochemistry and the kinetics of nutrition (Fyfe et al., 1983)

The generalized relationships between regolith, thickness of the weathering zone, climatic factors and water chemistry are shown in Figure 2.5. It has also been observed that the organic matter in the soil also tends to get depleted due to intense oxidation which often extends to the weathering front. This is generally accompanied by the depletion of some major elements such as P, N and K and soluble cations such as Na^+ and Ca^{2+} as well as anions Cl^-, PO_4^{3-} and NO_3^-. The natural vegetation of these infertile soils is therefore low and the essential trace elements needed for human and animal health become relatively sparse.

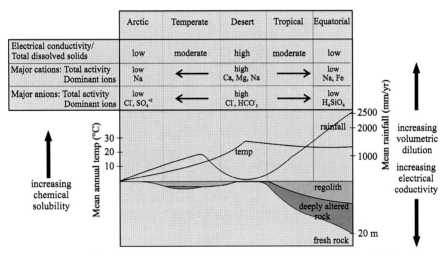

Fig. 2.5. Generalized relationships between regolith, thickness of weathering zone, climatic factors and water chemistry (Plant et al., 1996; Pedro, 1985)

Under tropical environmental conditions, element fractionation in the weathering of rocks and soil formation result in heterogeneous distribution of elements in terrains. Accordingly, areas of unique geochemical features develop with their inherent deficiency or excess of trace and major elements.

These special areas could also be termed as "geochemical provinces" and as shown in later chapters, they form an important facet of medical geology of tropical environments.

The bioavailability of essential and toxic elements plays a major role in the aetiology of environment-related diseases and this is influenced to a marked degree by the mobility or non-mobility of the elements concerned in the "rock-soil" environment. The laterites, so typical of the tropical environment, often function as "scavengers" of cations and anions and these, therefore, control the distribution of the elements in the 'geochemical provinces' to a marked degree. There are numerous external and internal factors that are responsible for the concentration and partitioning of elements in laterites.

The presence of active surfaces on which element adsorption and concentration takes place is an essential pre-requisite. Laterites contain abundant amorphous and poor crystalline constituents that show high adsorption rates. These were termed "scavengers" by Kühnel (1987).

The following are considered as common "scavengers" in soils and laterites, (i) amorphous silica (ii) Fe, Al and Mn-oxyhydroxides (iii) organic matter. Clay minerals such as smectite as well as phosphates, carbonates and some sulphides are also known to act as "scavengers". The high ability of these mineral phases to take up and fix elements from migrating solutions is mostly due to the presence of a very large surface area, which enhances chemical reactivity to a marked degree. The minute particle size and the amorphous nature is particularly responsible for the presence of the large surface area.

The ability of the "scavengers" to trap all available particles including water molecules, cations, or anions in governed by the need to compensate for the charge imbalances in the amorphous surface. Table 2.3 shows examples of some "scavengers" in laterites and soils present in the tropical environment.

Table 2.3. Examples of some "scavengers" in laterites and soils (Kühnel, 1987)

Scavenger	Collected cations (anions)	Possible products after recrystallization
Fe-oxyhydroxides	Ti, Cr, Al	magnetite, maghemite, hematite
	Cr, Al, Ni, Co, (F)	goethite and other FeOOH polymorphs
	Co	
	Cu	goethite + heterogenite
	(AsO$_4$)	delafossite, goethite + cuprite
Mn-oxyhydroxides		goethite + struvite
	Co, Ni	
	Ba, Zn, Al	todorokite, woodruffite
	Li, Al	psilomelane
	K, Na	lithiophorite
	Zn, Pb, Ag	cryptomelane, birnessite
	Ca, Mg	hetaerolite, quenselite
	Fe, Mg	marokite
Silica	(SiO$_4$)	bixbyite, jacobsite braunite
	Al, Ni, Na, K	
Al-oxyhydroxides +silica +magnesia	Cu	opal, jasper, quartz chryocolla
	Ni, Co, Cr	
	Ni	clay minerals
	Cu	smectites
	Ni, Fe, Al	talc, sepiolite, serpentinite
	Ni(CO)$_3$, (SO$_4$)	vermiculite pyroaurite pyroaurite, takovite

Tropical Weathering of Mineralized Terrains

Intense weathering in mineralized terrains in tropical environments brings about unique changes in the geochemistry and mineralogy in the terrain. People living in such areas are often subjected to a unique geochemistry. The largest reserves and resources of lateritic nickel for example, occur in developing countries in the tropics and sub-tropics.

The conditions necessary for the formation of laterite deposits resulting in the concentration of Ni, Co, Cr and Al are well known. Dissanayake (1984a) briefly discussed the main criteria for tropical weathering:

(a)   Climate: a humid tropical or sub-tropical climate with seasonal changes of rainfall and temperature promotes the laterization process. Due to the high intensity of the rainfall, metals are easily brought into solution and leached away. During the dry season, saturation of the metal ions in solution is followed by precipitation at specific physico-chemical environments resulting in concentrations of metals at various levels.

(b) Drainage: Good porosity of the rocks and easy drainage permit the circulation of leaching groundwaters. This enables the chemical processes involved in the weathering to achieve completion.

(c) Topography: Regions with strongly pronounced relief are not conducive to weathering, since erosion outweighs chemical weathering. A flat or nearly flat topography, especially on old land surfaces, enables optimum water table conditions to operate with the resulting accumulation of the products of chemical weathering.

(d) Time: Prolonged exposure to the above mentioned factors under stable tectonic conditions results in the near completion of the weathering processes and the presence of weathering products at different stages of evolution indicates the relative time of exposure.

(e) Parent rocks: Ultrabasic rocks such as peridotites are known to contain significant contents of Ni, Co and Cr (Turekian and Wedepohl, 1961) and the availability of such rocks covering large areas is a necessary criterion for the concentration of Ni, Co and Cr in the laterites.

Weathering Profiles

In lateritic terrains, weathering of different stages of evolution can be seen. Alteration of ultrabasic rocks into laterite involves complicated processes resulting in extreme changes of physical, chemical and mineralogical composition. Figure 2.6 illustrates two weathering profiles over serpentinites from Sri Lanka and Indonesia, respectively, the latter showing the mineralogical composition.

Weathering of Nickeliferous Serpentinites

In a typical cross-section, a distinct transition from the uppermost weathering products to the unaltered primary ultramafic rock can be seen. As shown in Figure 2.6 the lateritic iron ore or the cap can be seen on the top.

In places where decayed plant material is absent, the duricrust can be seen as a solid brown cap-like covering composed mainly of secondary iron-oxides and hydroxides. Iron has been leached out and re-precipitated by solution activity. On a nickeliferous serpentinite in Sri Lanka, Dissanayake (1984a) observed that the duricrust is distributed as bare patches generally devoid of vegetation. The iron-bearing material generally takes the form of massive rounded reddish brown nodules of a concretionary nature. It is common to find the surface strewn with such iron-ore concretions of varying sizes.

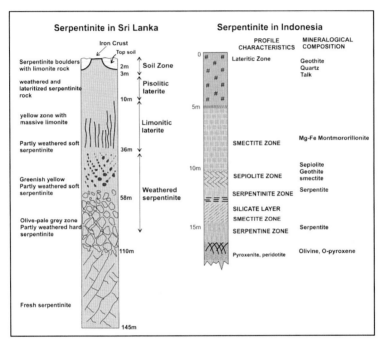

Fig. 2.6. Weathering profiles over serpentinite in Sri Lanka (Dissanayake, 1984a) and Indonesia (Kühnel et al., 1978)

Below the lateritic cap could be seen the remnants of highly weathered serpentinized ultramafic rock retaining in most cases the original reticulate or banded structure. A conspicuous feature in this zone is the occurrence of small black grains of magnetite and chrome spinels. This feature has been observed in a number of case histories (Sri Lanka: Dissanayake and Van Riel, 1976; India: Ziauddin and Roy, 1970; Thailand: Pungrassami, 1970; Cuba: Fisher and Dressel, 1959). It is also common to find the weathered ultrabasic rock assuming varying shades of green. It is of interest to note that Brindley and Hang (1973) had observed a close correlation of the intensity of green colour as determined by the Munsell colour chart, with the nickel content. This feature is worthy of consideration in nickel prospecting. Generally, the fragments and blocks become larger and more abundant downwards. The concentration of chrome-spinels, which are resistant to weathering increase upwards in the weathered zone. It is the material of this zone which needs to be analyzed carefully for nickel.

The next layer below is generally characterized by the occurrence of green coloured clays-sometimes called nontronites. While the original reticulate and banded structure may remain, the chrome-spinels and magnetite may be inconspicuous in this zone. Another important feature is the occasional occurrence of secondary silica in the form of chert, agate, opal and chalcedony. These have resulted from the leaching of silica by descending solutions and in the presence of significant concentrations of nickel assume a green colour.

Just above the fresh parent ultramafic rock is the partly weathered material. A noteworthy feature of this zone is the occurrence of carbonates in the form of calcite/magnesite. Their presence imparts a white appearance to the rock and pockets of shining white carbonate aggregates may indeed be found as in the case of the nickeliferous serpentinite of Sri Lanka.

The above-mentioned weathering profile, even though highly generalized, can be observed over ultramafic rocks in humid tropical terrains. The thickness of the various zones, however, varies widely depending on the locations. In some instances as in the case of the Moa serpentinized ultramafic mass of Cuba (Linchenat and Shirokova, 1964), the weathering profile may attain depths greater than 50 metres. It should be noted that nickel is concentrated in the weathered serpentinite zone up to about 10 times the parent rock value and generally reaches a peak close to the upper boundary zone. This feature of nickel being concentrated at the base of the weathering profile is very characteristic of most nickeliferous laterites and should be noted carefully. It is also of interest to note that cobalt is concentrated slightly above the zone where nickel is enriched.

Formation of Secondary Minerals

During lateritic weathering the formation of new and secondary minerals constitutes an important phase. It is therefore of great importance to locate and identify the neo-mineralization. This helps in the study of chemistry of the weathering profile in terms of partitioning of elements and also in selecting the proper metallurgical procedures for extraction. With the development of optical microscopy, electron microscopy, X-ray diffraction techniques, the cryptocrystalline characters of these secondary mineral phases have been studied in detail (e.g.: Kühnel et al., 1978; Chukrov, 1975; Esson and Carlos, 1978; Brindley, 1978). The surface area of such crystallites is extremely large and this determines the reaction rates during element concentration.

Brindley and Maksimovic (1974) and Brindley (1978) have systematically classified the hydrous nickel-containing silicates, ubiquitously present in the weathering profiles associated with nickeliferous laterites as follows:

1 : 1 layer type minerals	chrysotile-pecoraite lizardite-nopouite berthierine-brindleyite
2 : 1 layer type minerals	talc-willemsite kerolite-pinielite clinochlore-nimite sepiolite-falcondite

The term garnierite, following the recommendations of Pecora et al. (1949) and Faust (1966) is now accepted as a general term used widely when more descriptions cannot be given. Apart from this various other new mineral names such as nontronite, vermiculite and schuchardite are now in use. For detailed descriptions of the nomenclature of hydrous nickel silicates, the reader is referred to the work of Brindley and his co-workers (Brindley and Hang, 1973; Brindley and Maksimovic, 1974; Brindley and Wan, 1975; Brindley, 1978). The distribution of some of these secondary mineral phases in a weathering profile is illustrated in Figure 2.6.

Chemistry of Weathering of Ultra-basic Rocks

The change from a fresh ultrabasic rock to a fully formed laterite is a result of extensive geochemical remobilization and can be exemplified by the comparison of the chemical and mineralogical compositions of an average peridotite altered subsequently to a serpentinite and a laterite. The data in Table 2.4 obtained from the same bore-hole illustrate the degree to which chemical weathering alters a rock with the resulting remobilization and concentration of elements. Among the most striking features observed is the leaching out of magnesia and silica. It is of interest to note that these two oxides constitute nearly 80-90% of the original unweathered rock but in the end-product only about 1-2% remains - an indication of the extreme leaching and accompanying remobilization. The descending meteoric waters contain carbonic and organic acids and these are responsible for the leaching out of some of the constituents of the original rock. The extreme leaching and concentration of the elements during the laterization of the ultrabasic rocks is illustrated in Figure 2.7. Alumina and also oxides of titanium are somewhat stable components of the parent rock and are generally used in the material balance composition.

Table 2.4. Chemical composition and mineralogical composition of average peridotite, serpentinite and laterite (after Kühnel et al., 1978; * from the same borehole; tr- traces)

Chemical Composition	Average Peridotite*	Serpentinite	Laterite
SiO_2	43.9	43.2	1.84
TiO_2	0.8	tr	0.14
Al_2O_3	4.0	1.7	9.00
Fe_2O_3	2.5	10.6	67.40
FeO	9.9		
NiO	0.2	2.8	1.50
MnO	0.1	0.2	0.80
MgO	34.3	32.4	0.80
CaO	3.5	tr	tr
Na_2O	0.6	tr	tr
K_2O	0.2	tr	tr
H_2O	-	13.9	14.22
Cr_2O_3	0.3	0.2	3.77
S	tr	tr	tr
Cl	tr	tr	tr
Mineralogical Composition	Olivine	Serpentine	Goethite
	Pyroxene	Spinel	Hematite
	Spinel	Hematite	Serpentine
	Sulphide	Goethite	Spinel
		Magnesite	Clay Mineral
		Talc	
		Clay mineral	Silica

Geochemistry of the Tropical Environment 35

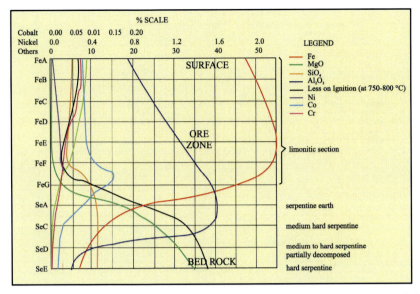

Fig. 2.7. Variation of elements with depth in a weathering profile over serpentinite (after de Vletter, 1978)

HYDROGEOCHEMISTRY OF THE TROPICAL ENVIRONMENT

The distribution of the different chemical species in the various components of the hydrological cycle has a marked influence on the medical geology of tropical lands, particularly in relation to groundwater related diseases.

Figure 2.8 illustrates the components of the shallow groundwater cycle that are sensitive to change. Several geochemical processes could take place in the groundwater cycle that could markedly alter the chemistry of the water. From among these the following processes are considered important:

(a) evaporation and evapotranspiration,
(b) selective uptake of ions by vegetation,
(c) decay of organic matter,
(d) weathering and dissolution of minerals,
(e) precipitation of minerals,
(f) mixing of different water qualities and
(g) anthropogenic activities.

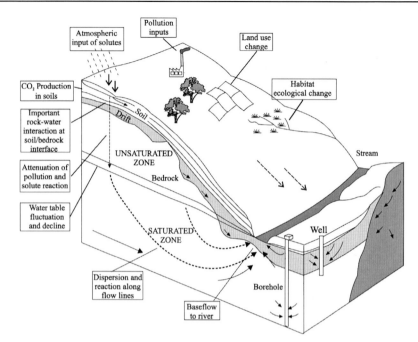

Fig. 2.8. Components of the shallow groundwater cycle showing elements that are sensitive to environmental change (Edmunds, 1996)

A schematic overview of the processes that affect water quality in the hydrological cycle is shown in Figure 2.9. Table 2.5 shows the processes which are important as sources of different ions and processes that may limit their concentration in fresh water. Trescases (1992) has shown that chemical elements are more likely to undergo pronounced fractionation in tropical environments than in temperate regions. The process of chemical separation, however, depends on their chemical mobility and the nature of the local environment. On account of this, tropical environments are far more likely to have areas with the potential for deficiency or toxicity conditions (Plant et al., 2000). The relative mobility of some essential and potentially toxic elements in different surface conditions is shown in Figure 2.10.

Table 2.5. Processes which are important as sources of different ions and processes that may limit their concentration ions in fresh water (after Appelo and Postma, 1994)

Elements	Process	Concentration Limits
Na^+	Dissolution, cation exchange in coastal aquifers	Kinetics of silicate weathering
K^+	Dissolution, adsorption, decomposition	Solubility of clay minerals, vegetation uptake
Mg^{2+}	Dissolution	Solubility of clay minerals
Ca^{2+}	Dissolution	Solubility of calcite
Cl^-	Evapotranspiration	None
HCO_3^-	Soil CO_2 pressure, weathering	Organic matter decomposition
SO_4^{2-}	Dissolution, oxidation	Removal by reduction
NO_3^-	Oxidation	Uptake, removal by reduction
Si	Dissolution, adsorption	Chert, chalcedony solubility
Fe	Reduction	Redox-potential, Fe^{3+} solubility, siderite, sulfide
PO_4^{3-}	Dissolution	Solubility of apatite, Fe, Al phosphates. Biological uptake

In an interesting study on the biogeohydrodynamics in a forested humid tropical environment in south Cameroon, Braun et al. (2002), perhaps for the first time world-wide, attempted to combine different approaches in hydrology, biogeochemistry, mineralogy, crystallography, microbiology, geophysics and pedology. One of the major results they obtained is the essential role played by the vegetation and soil organic matter in the fractionation, distribution or storage of the chemical elements in the humid tropical environment. They noted a strong geochemical contrast between the different groundwater zones-namely, flooded areas, hill slope lateritic profiles, weathering interface between the saprolite and the basement rocks and the swampy areas. Humic substances were particularly important in the weathering budget of elements usually considered as immobile in the surficial cycle, e.g.: Al, Th, Zr, and Fe.

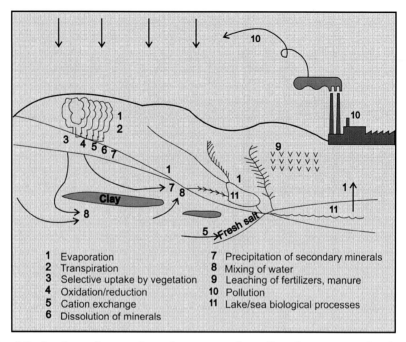

Fig. 2.9. A schematic overview of processes that affect the water quality in the hydrological cycle (Appelo and Postma, 1994)

The relatively high content of the oxygen containing functional groups such as CO_2H, phenolic and alcoholic OH, ketonic and quinonoid C=O provide mechanisms for the easy interaction of organic soil and water constituents (Schnitzer, 1978). It is through the acid functional groups that these materials are thought to interact with metal ions to form metal-organic associations of widely differing chemical and biological characteristics.

Kerndorff and Schnitzer (1980) investigated the sorption of humic acid of metals from an aqueous solution containing Hg (II), Fe (III), Pb, Cu, Al, Ni, Cr (III), Cd, Zn, Co and Mn with special emphasis on effects of pH, metal concentration and humic acid concentration. They observed that the sorption efficiency tended to increase with rise in pH, decrease in metal concentration and increase in humic acid contents of the equilibrating solution. The orders of sorption were as follows;

pH 2.4: Hg > Fe > Pb > Cu = Al > Ni > Cr = Zn = Cd = Co = Mn
pH 3.7: Hg = Fe > Al > Pb > Cu > Cr > Cd = Zn = Ni >= Mn
pH 4.7: Hg = Fe = Pb = Cu = Al = Cr > Cd > Ni = Zn > Co > Mn
pH 4.7: Hg = Fe = Pb = Al = Cr = Cu > Cd > Zn > Ni > Co > Mn

Geochemistry of the Tropical Environment

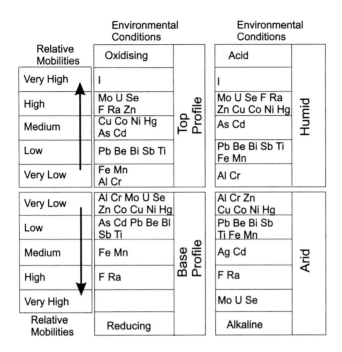

Fig. 2.10. The relative mobility of some essential and potentially toxic elements in different surface conditions (Plant et al., 2000)

The abundance and distribution of organic matter therefore plays a very significant role in a geochemical province where the element distribution shows a unique pattern. The water quality of groundwater in organic-rich soils will show special geochemical signatures as against groundwater in soil notably absent in organic matter.

In developing countries, many of which are in the equatorial humid tropical belt, the chemistry of the groundwater is of special importance in relation to medical geochemistry. The ground and surface water is often directly used as drinking water and the link between groundwater chemistry and health is obvious.

The groundwater derives its solutes from contact with various solids, liquids and gases as it finds its way from the recharge to the discharge area. The chemical composition of the rocks, mineral and soils through which the groundwater flows causes very large variations of the chemistry of the groundwater. In rural areas of the developing countries of the tropics, particularly in areas where contamination due to industrial emission is minimal, the chemistry of the rocks, minerals and soils of a terrain is of paramount importance in the health of the indigenous people. Table 2.6 shows the concentration ranges of dissolved inorganic constituents in groundwater. These constituents are all ions except for aqueous silica. Minor and trace constituents of the water are far greater in number and anomalous situations occur when some of the trace elements shown in Table 2.6 reach very high concentrations and these would undoubtedly have an impact on the health of the people living in that area. Excess nitrates caused by heavy inputs of nitrogen–bearing fertilizers, heavy metals in acid soils around mineral deposits are examples. Iron is also known to show elevated concentrations around lakes, wet lands and in contaminant plumes (Kehew, 2001).

In the tropical environment, the water quality could be conveniently displayed in a piper diagram. It consists of two triangles one for cations and one for anions and a centrally located diamond-shaped figure. The concentrations of Ca, Mg and Na+K are on the three axes of the cation triangle. Similarly, Cl^-, SO_4^{2-} and $CO_2^- + HCO_3^-$ are plotted in the anion triangle. The water analysis depicted as a point in the two triangles is now projected onto the rectangle (Fig. 2.11).

The value of a piper diagram is further enhanced by its ability to classify the water types based on the water analyses plotted in the ion triangles. The water types also termed chemical facies, are illustrated in Figure 2.11. Medical geology has greatly benefited by the use of these Piper plots since geochemical provinces with their inherent water quality characteristics can be linked to problems of human and animal health. Such classifications of water types are particularly relevant to studies of the medical geology of cardiovascular diseases, where water hardness is a major factor.

In Bangladesh, a tropical humid country where leaching of soils is intense throughout the country, variability in climate causes variability in weathering rates. Islam et al. (2000) noted that spatial variability exists in the rate of mobilization/retention of both major and trace elements suggesting the importance of local controls. These authors noted that lakes and reservoirs are also generally enriched in all measured elements (Fig. 2.12), sometimes independently of the degree of depletion or enrichment of the soil profile. The study of Islam et al. (2000) in a tropical environment devastated by natural hazards, including adverse medical geochemistry (e.g. As, Al) is interesting from the point of view that it is not only the groundwater that had received high inputs of major and trace elements, but surface water as well, compounding the drinking water problem of Bangladesh.

Table 2.6. Concentration ranges of dissolved inorganic constituents in groundwater (Freeze and Cherry, 1979; Davis and De Wiest, 1966)

Major constituents (greater than 5 mg/L)	
Bicarbonate	Silica
Calcium	Sodium
Chloride	Sulfate
Magnesium	
Minor constituents (0.01-10.0 mg/L)	
Boron	Nitrate
Carbonate	Potassium
Fluoride	Strontium
Iron	
Trace constituents (less than 0.1 mg/L)	
Aluminium	Molybdenum
Antimony	Nickel
Arsenic	Niobium
Barium	Phosphate
Beryllium	Platinum
Bismuth	Radium
Bromide	Rubidium
Cadmium	Ruthenium
Cerium	Scandium
Cesium	Selenium
Chromium	Silver
Cobalt	Thallium
Copper	Thorium
Gallium	Tin
Germanium	Titanium
Gold	Tungsten
Indium	Uranium
Iodide	Vanadium
Lanthanum	Ytterbium
Lead	Yttrium
Lithium	Zinc
Manganese	Zirconium

The chemical changes that occur in groundwater flow systems influence the hydrogeochemical anomalies that may be present locally and which may eventually have a bearing on the human and animal population. It has been observed (Chebotarev, 1955) that groundwater tends to evolve chemically in long flow systems resulting in more concentrated solutions. As shown by Kehew (2000), this sequence is as follows:

$$HCO_3^- \rightarrow HCO_3^- + SO_4^{2-} \rightarrow SO_4^{2-} + HCO_3^- \rightarrow SO_4^{2-} + Cl^- \rightarrow Cl^- + SO_4^{2-} \rightarrow Cl^-$$

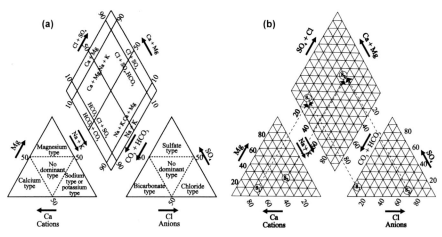

Fig. 2.11. (a) Classification of water types using the Piper trilinear diagram, (b) water analyses plotted on Piper trilinear diagram

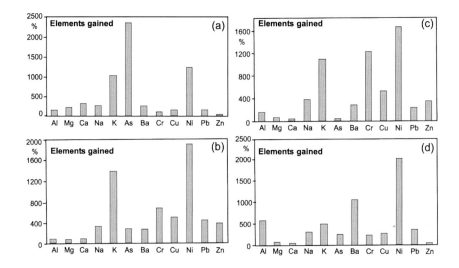

Fig. 2.12. Percent increase in dissolved elements in water of tropical environment in comparison to worldwide typical values: Rajarampur, b. Shamta, c. Mainamoti, d. Andulia (after Islam et al., 2000)

Tóth (1963) classified the groundwater flow systems as local, intermediate and regional flows systems, the size of each of them being governed by the topography in the drainage basin relative to the depth of the flow system. Figures 2.13 and 2.14 illustrate the distribution and hydrochemical facies of the different types of groundwater flow systems. From the point of view of medical geochemistry, the recharge and discharge areas are of significance in view of the hydrogeochemical anomalies that may be present at such localities, bearing in mind that the chemical composition of groundwater in flow paths that cross multiple rock types can be highly variable. Figures 2.15 shows such a scenario in a tropical environment in which the base rock mineralogy, soils, morphology and the ground water flow regime combine to form geochemical anomalies.

The presence of mineralization can also affect the groundwater quality of a terrain and this may have an impact on the health of the population living in such terrains. It is known that at depth, groundwater can transport ore and host elements both laterally and vertically. Exploration geochemists therefore use groundwater chemistry in regions where mineral deposits are deeply buried by a thick overburden as a tool in mineral exploration. The geochemical signatures from such buried mineralization are very often not amenable to most exploration techniques. The interaction of the groundwater within a body of mineralization brings about a change in the quality of groundwater chemistry. Groundwaters can therefore be used as pathfinders to concealed ore deposits. In an article published by the Australian Institute of Geoscientists (AIG, 2003), over 5500 groundwaters across mineral provinces of Australia have been sampled and the method applied.

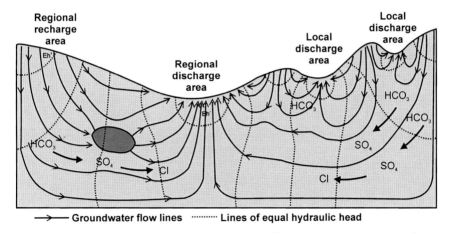

Fig. 2.13. Hydrochemical facies related to type of flow system in hypothetical regionally unconfined basin (Tóth, 1999)

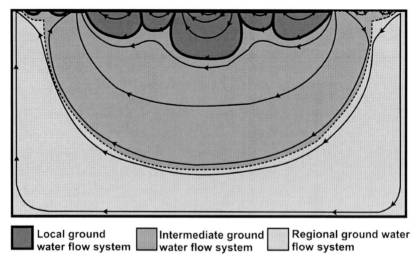

Fig. 2.14. Distribution of local, intermediate and regional groundwater flow systems in a drainage basin (Tóth, 1963, reproduced with kind permission from American Geophysical Union)

The field measurements include pH, Eh, salinity, temperature, reduced Fe, and trace elements. Geochemical anomalies observed from such geochemical exploration programmes for concealed mineralization can also be effectively used in projects aimed at delineating areas of anomalous trace element contents that could cause increased incidence of certain diseases. Areas underlain by granitic rock bodies may, for example, contain higher fluoride and rare-earth element concentrations and these may lead to diseases such as dental and skeletal fluorosis and endomyocardial fibrosis, among others. Further, the abundance of total dissolved solids (TDS), Ca and Mg carbonates, salinity etc in the water caused by the presence of rock types that enhance concentrations of such chemical components, may also have health implications.

A good understanding of the geology and the lithology of a terrain, its surface and groundwater chemistry, groundwater flow systems, their recharge and discharge areas and geomorphological aspects will be a useful pre–requisite for any study concerning the epidemiology of an endemic disease such as those prevalent in tropical environments.

Geochemistry of the Tropical Environment

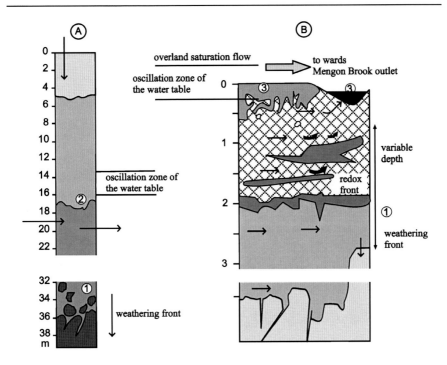

Fig. 2.15. Reference profile of a hill top (A) swamp (B) and location of the sampling waters. (1) Clear waters of the weathering front (2) clear water of the upper fringe of the hillside groundwater (3) coloured water of the surface groundwater and of the Mengong Brook (Braun et al., 2002)

CHAPTER 3

BIOAVAILABILITY OF TRACE ELEMENTS AND RISK ASSESSMENT

In the field of medical geology, bioavailability and risk assessments are two important aspects that need serious consideration. The mere presence of a toxic trace element or a species in a specific locality for example, does not necessarily imply a serious health hazard. The nature of the species concerned and the bioavailability of the species are the main factors that influence the impact of an environmental chemical species on human and animal health. The bioaccumulation of the species concerned is an important step in the food chain and the processes controlling it needs to be well understood.

BIOACCUMULATION

In a very general sense, bioaccumulation could be defined as "the process by which organisms absorb chemicals or elements directly from the environment". It should however be noted that the term "bioaccumulation" has to be specified by quantitative data, comparing concentrations of a compartment in relation to another (e.g. plant vs. soil parts, adsorbed amount vs. dissolved amount) (Streit, 1992). The selective concentration of elements is inherent in any life process and the process of bioaccumulation therefore leads to indications of pollution of the environment. The chemistry of the elements and the biochemical structure of the biological species are the main factors that influence the bioaccumulation process.

The passage of metal ions into the plants through the soil solution and root cells depends on a number of factors such as redox potential, pH, interaction with ligands and properties of soil matrix such as cation exchange capacity. The metal speciation in soil solution and on soil surfaces and metal uptake by the root is illustrated in Figure 3.1. The root cell membranes are highly selective to trace elements. The phospholipid based membranes are highly impermeable to ions or (large) polar molecules, whereas non-polar molecules (such as O_2) pass rapidly. The electric potentials between membrane separated compartments, known to correlate with pH gradients play an important role in soil solution-plant pathways (Streit and Stumm, 1993). The mechanism of element passage through the cell membranes is complex and very little is still known about the actual selective process.

BIOAVAILABILITY

The term bioavailability has been defined in a number of ways. In a very general way, one could define bioavailability as "the extent to which a substance can be absorbed by a living organism and can cause an adverse physiological or toxicological response" (NEFESC, 2000). For environmental risk assessments which involve soil and sediment, the above definition implicitly includes the medium in which it occurs to become available for absorption.

Other definitions of bioavailability depend on the scientific discipline which requires such a definition. For example:

(i) Environmental bioavailability: Physiologically driven uptake process (Peijnenburg et al., 1997)

(ii) Toxicological bioavailability: The fraction of the total available dose absorbed by an organism which is distributed by the systematic circulation and ultimately presented to the receptor or sites of toxic action (Landrum and Hayton, 1992)

(iii) Bioremediation bioavailability: The extent to which a contaminant is available for biological conversion (Juhasz et al., 2003).

Other terms such as pharmacological bioavailability, phytobioavailability and bioaccessibility have also been used (Landrum and Hayton, 1992).

Bioavailability of Trace Elements

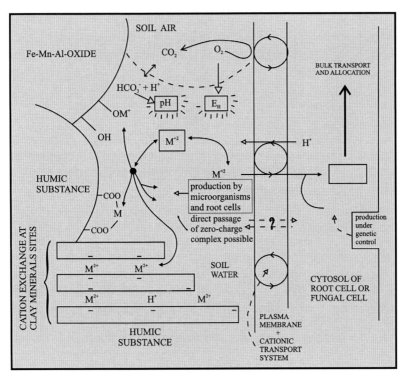

Fig. 3.1. Highly schematic diagram to show metal speciation in soil solution and on soil surfaces and root metal ion uptake, assuming a divalent metal cation (M^{2+}) and a monovalent soluble external ligand. The uptake rate into the cell is determined (in a given individual) by the number of metal ions bound to receptor sites on membrane transport molecules, e.g. metal ion ATPases. At equilibrium the amount of metal bound to these sites and therefore the transport rate, would be directly related to the soil metal ion concentration (Streit and Stumm, 1993)

The three main factors that control bioavailability are:

(a) there has to be an opportunity for the receptor (or organism) to be in contact with the matrix in which the contaminant is found.

(b) the contaminant must be potentially available at least in part.

(c) the receptor (or organism) must be able to absorb or assimilate the potentially available fraction.

If these three requirements are not fulfilled, a contaminant is not bioavailable (Juhasz et al., 2003). In the case of human health risk assessment, two types of bioavailability, namely absolute bioavailability and relative bioavailability are measured. Absolute bioavailability is the fraction or percentage of a compound which is ingested, or applied on the skin surface that is actually absorbed and reaches the systemic circulation, and is defined as:

$$\text{Absolute Bioavailability} = \frac{\text{Absorbed dose}}{\text{Administered}} \times 100$$

Since toxicity parameters are based on an administered dose rather than an absorbed dose, absolute bioavailability in often not determined in the case of human health risk assessments.

Relative bioavailability on the other hand, is a measure of the extent of absorption among two or more forms of the same chemical, different media (e.g. food, soil, water) or different doses. Since matrix effects can substantially decrease the bioavailability of a soil or sediment bound metal compared to the form of metal or dosing medium, relative bioavailability is more important in environmental studies. It is defined as:

$$\text{Relative Bioavailability} = \frac{\text{Absorbed fraction from soil}}{\text{Adsorbed fraction from dosing medium used in toxicity study}}$$

The relative bioavailability thus expressed is termed the relative absorption fraction (RAF). In the case of trace elements, bioavailability varies considerably depending on factors such as food source, oral intake, chemical form or species, nutritional state (deficiency vs. excess), age, gender, physiological state, pathological conditions and interaction with other substances.

The determination of bioavailability is carried out in a number of ways. Among these are:
(a) microbial tests
(b) soil invertebrate tests
(c) amphibian tests
(d) plant tests
(e) tests using higher organisms

These tests use the measurement of a toxic or a mutagenic response, inhibition of a metabolic function, changes in microbial population structure, mortality or malformations, accumulation of chemical species in organs or the blood streams, and dissolution of contaminants after extraction of particular mineral phases (Juhasz et al., 2003).

In the soil environment several processes affect the mobility of metals. These are: (a) dissolution and precipitation (b) sorption (c) ion exchange and (d) oxidation-reduction reactions. In such an environment, the mobility and hence the bioavailability of metals is reduced by conditions that promote precipitation or sorption. The most bioavailable metals are those that form weak outer complexes with inorganic materials such as clay, iron and manganese oxides or organic soil matter or those that form complexes with ligands while in solution. These are therefore not sorbed. On the otherhand, metals that tend to form inner sphere complexes do not desorb easily and hence less bioavailable. Table 3.1 shows the relative mobility of some metals in soils. The bioavailability of these metals can be inferred from these mobilities.

In the case of sediments, metals may be to a large extent incorporated into the structure of the mineral itself and hence may not be easily bioavailable. Only the remaining metals which are adsorbed or complexed may become bioavailable (Table 3.2).

RISK ASSESSMENT

Risk assessment is a method which assesses the actual or potential adverse effects of contaminants to plants and animals and which focuses on the damage that has been or will be done by contaminants. The information obtained from risk assessment helps one to identify populations or areas that are likely to be adversely affected by soil, water or air contamination.

The technique of risk assessment (WHO, 2002) is based on a causal stress-response model where a contaminant is transported from its source to the receptor (humans, animals, plants) through a definite pathway. The main components of a risk assessment are:

a. Identification of the problem
b. Characterization of the receptor
c. Assessment of the exposure
d. Assessment of the toxicity
e. Characterization of the risk

Table 3.1. Relative mobility of selected metals in soil (Hayes and Traina, 1998)

Metal	Most common Oxidation states in soil[a]	Predominant Forms and Distribution in soil systems	Mobility
Arsenic	III	Oxyanion: sorbs more weakly than As(V) to metal oxides and only at higher pH	Moderate
Arsenic	V	Oxyanion: sorbs strongly to metal oxides; forms relatively insoluble precipitates with iron	Low
Cadmium	II	Cation: sorbs moderately to metal oxides and clays; forms insoluble carbonate and sulphide precipitates	Low to moderate
Chromium	III	Cation: sorbs strongly to metal oxides and clays; forms insoluble metal oxide precipitates	Low
Chromium	VI	Oxyanion: sorbs moderately to metal oxides at low pH, weaker sorption at high pH	Moderate to High
Lead	II (IV)	Cation: sorbs strongly to humus, metal oxides, and clays; forms insoluble metal oxides and sulphides; forms soluble complexes at high pH	Low
Mercury	II (O-I)	Cation: sorbs moderately to metal oxides and clays at high pH; relatively high hydroxide solubility; forms volatile organic compounds	Low
Nickel	II (III)	Cation: sorbs strongly to humus, metal oxides, and clays; forms insoluble metal oxides and sulphides; forms soluble complexes at high pH	Low

[a] Possible, but less common, oxidation states in soil systems are shown in parentheses.

Risk assessment involves ecological and human risk assessment. The former, in view of the large number of receptors involved is far more complicated than the latter. Human risk assessment however, involves serious ethical considerations and in vivo studies using many animal models are therefore more often used.

Table 3.2. Dominant absorbed or complexed phases of metals in oxic and anoxic sediments (from Brown and Neff, 1993) [-CO$_3$ = carbonates; FeO = iron oxyhydroxides; Fe/MnO = iron and manganese oxyhydroxides; OM = organic matter; S = sulfides (dominant species given); TBL-Cl-OH-CO$_3$ and S = tributyltin chloride, hydroxide, carbonate, and sulfide]

Metal	Association in Oxic Sediments	Association in Anoxic Sediments
Arsenic	AsO$_4^{3-}$ - Fe/MnO	As$_2$(SO$_3$)$_2$, As$_2$S$_3$, FeAsS
Cadmium	Fe/MnO, OM/S, -CO$_3$	CdS
Chromium	OM, FeO	OM, Cr(OH)$_3$
Copper	OM, Fe/MnO	Cu$_2$S, CuS, FeCuS
Lead	Fe/MnO	PbS
Mercury	OM	HgS, OM
Nickel	Fe/MnO	OM/NiS, organic thiols
Tin[a]	TBL, –Cl, –OH, -CO$_3$	TBL-S, OH, -CO$_3$
Zinc	Fe/MnO, OM	ZnOM/S

(a) Only butyl tins are considered

The ecological risk assessment involves the uptake of metals by plant and animals from soils, sediments and water. The processes that occur within the food chain are clearly complex and ecological risk assessment needs to consider several aspects of metal uptake. Among these are the bioavailability of metals by plants and animals from soils, sediments and water by contact with external surfaces, ingestion of contaminated soil, sediment or water and inhalation of vapour phase metals or airborne particles. Food is another source for the bioavailability of metals in animals. Based on these factors, bioavailability is evaluated by estimating:

(a) available fraction of metals present in the environmental media (e.g. sediment or soil)
(b) bioaccumulation directly from environmental media
(c) uptake from ingestion of food

It has now become clear that bioavailability must necessarily become an integral part of the method of risk assessment. Risk-based approaches for the assessment of contaminated sites using total soil metal concentrations rather than the bioavailable metal concentrations, for example, may lead to erroneous conclusions. A good understanding of the bioavailability of the contaminant concerned is therefore a definite pre-requisite for all risk assessment procedures. Of special importance is the need to understand the factors that affect bioavailability, human health implications, plant uptake and ecosystem health of metals. The field of medical geology clearly recognizes the impact of studies of bioavailability on epidemiology.

ASPECTS OF EPIDEMIOLOGY IN MEDICAL GEOLOGY

Epidemiology is that branch of medicine that investigates the frequency and geographic distribution of diseases in a defined human population for the purpose of establishing programmes to prevent and control their development and spread. Geochemistry plays a major role in the field of epidemiology, particularly in diseases of tropical environments in developing countries, where large human and animal populations live in intimate contact with soil, water and plants of their habitat. The link between geology and epidemiology becomes direct in such cases. The term prevalence, in relation to a disease, indicates the proportion of the population that has a particular disease, at a specific time. Incidence measures the frequency of new cases of the disease.

Epidemiology involves the investigations of the origins of a specific disease, also termed aetiology, and to develop and test the hypotheses. It also attempts to discover the likelihood of a population being exposed to the disease concerned and identifies risks in terms of probability statements and studies trends over time to make projections for the future. New risk factors are discovered and studied in detail. Epidemiology deals extensively with mortality or death rates and morbidity and the rate of incidence of the disease concerned.

One of the terms used frequently in epidemiological investigations is "risk factor". The risk factor could be associated with the natural soil, water, plant, air and biological environment or an anthropogenically originated medium. The risk factor is any characteristic or condition that may occur with greater frequency in people with a disease than it does in people known to be free from that disease. Based on the information on the risk factor, the chance of increased spread of the disease can be studied and remedial measures recommended. The general term "risk" is often classified in chemical epidemiology as relative and absolute risk. The former is used to denote the ratio of incidence or prevalence in the exposed group to that of the unexposed groups. The absolute risk is the chance of a person to develop a disease.

The epidemiologists are often confronted with 'competing risks' which in fact are other sets of risk factors than can cause the condition of interest and which coexist with the set of factors of interest. These are known as "red herring" cases in out break investigations.

CAUSATION AND CORRELATION

These two terms are very frequently used in both medical geology and epidemiology. The term 'association' is also used in conjunction with correlation if two variables appear to be related by a mathematical relationship, indicating that a change of one appears to be related to the change in the other. Causation is used in epidemiology when the following conditions are satisfied;

(a) A dose-response relationship exists between the condition and the disease.
(b) The prevalence or the incidence of the disease is reduced when the condition under study is removed.
(c) The condition precedes the disease.
(d) A cause and effect relationship is physiologically plausible.

It is however, most important to note that a correlation does not always imply a causal relationship even though a correlation is necessary for a causal relationship. A correlation can be of two types, namely a negative correlation and a positive correlation. In the former, the magnitude of one variable moves in the opposite direction to the other associated variable. Here the correlation coefficient is negative. If the relationship is definitely causal, then the higher levels of the risk factor are protective against the outcome. In the case of a positive correlation, the two variables change in the same direction and the correlation coefficient is positive. Here, the higher the levels of the risk factor, the higher are the outcome. Correlation however measures only linear association. The fact that a statistical correlation alone does not prove causation may be due to the presence of other contributing or confounding factors. This provides misleading evidence and lead researchers to find an association for the wrong reason.

HOMEOSTASIS IN MEDICAL GEOLOGY

The geochemical cycles of the elements trace their pathways in the lithosphere, atmosphere, hydrosphere and biosphere. When these elements (both essential and toxic) enter the food chain and the human body, a process termed 'homeostasis' operates maintaining equilibrium within the constituent cells.

Homeostasis is the maintenance of equilibrium, or constant conditions, in a biological system by means of automatic mechanisms that counteract influences tending towards disequilibrium. This concept was developed by the French physiologist Claude Bernard in the middle of the nineteenth century and is presently considered as one of the most fundamental concepts in modern biology. The term homeostasis was introduced by Cannon in 1932 and was defined as "a condition which may vary, but remains relatively constant" (Clancy and McVicar, 1995).

In the human body all the systems are involved in homeostasis with greater involvement of endocrine, nervous, respiratory and renal systems. In the case of an imbalance, these regulating systems work towards restoring the optimum conditions. This is done by a process termed 'negative feedback' in which a deviation from the normal level is detected and the restoration process initiated.

The process 'contact inhibition' is an example of homeostasis where cell division in a population of cells ceases when they are too numerous or touch each other. A chemical 'messenger' is thought of as passing from cell to cell thereby causing the inhibition. Cancer cells on the other hand are propagated without any inhibition and hence they have lost the mechanism of homeostasis.

In the case of trace elements in humans, the principle of homeostasis can be applied in order to trace the pathways of the elements concerned within the human body. Even though the study of the elemental pathways within the human body lies in the domain of human physiology, the elemental pathway is part of the natural cycle of the element and hence it overlaps with the domain of geochemistry.

The World Health Organization (WHO, 2002) in their report on "Principles and Methods for the Assessment of risk from Essential Trace Elements (ETE)" classified the ETEs as comprising of three groups. These are

(a) cations (Zn, Fe, Cu, Mn, and Cr), (b) anions (those of Mo, I, Se) and (c) those forming bioinorganic complexes (e.g. Co). For each category the body has evolved specific mechanisms for the acquisition and retention, storage and excretion of the various elements.

Cationic ETEs: via gastrointestinal tract and liver. Homeostasis regulates uptake and transfer of the metal by the gut (e.g. Fe, Zn, and Cu). Copper for instance shows a decrease in absorption from 75% at 0.4 mg/day to 12% at 7.5 mg/day even though the total amount absorbed increases.

Anionic ETEs: These are more water soluble and less reactive with N, S, P, O, and OH groups than are cationic ETEs. They are absorbed very efficiently (>70%). Homeostasis is managed by manipulation of oxidation and methylation states. Total body burden is regulated by renal excretion.

ETEs forming bioinorganic complexes: Since cobalt can compete with many other cationic ETEs effectively, it is possible that cobalamin forms to avoid such problems (WHO, 2002).

The WHO has applied the homeostatic model in human health risk assessment for ETEs as follows:

> Homeostatic mechanisms should be identified for the selected ETE.
> Variation of the population's homeostatic adaptation must be considered.
> There is a "zone of safe and adequate exposure for each defined age and gender groups "for all ETEs – a zone compatible with good health. This is the acceptable range of oral intake (AROI).
> All appropriate scientific disciplines must be involved in developing an AROI.
> Data on toxicity and deficiency should receive equal critical evaluation.
> Bioavailability should be considered in assessing the effects of deficiency and toxicity.
> Nutrient interactions should be considered when known.
> Chemical species and the route and duration of exposure should be fully described.
> Biological end-points used to define the lower recommended dietary allowances (RDA) and upper (toxic) boundaries of the AROI should ideally similar degrees of functional significance. This is particularly relevant where there is a potentially narrow AROI as a

result of one end-point being of negligible clinical significance. All appropriate data should be used to determine the dose-response curve for establishing the boundaries of the AROI.

These are several steps involved in the application of the principles for the assessment of risk from essential trace elements. These are:

Step 1: Data selection. This develops a data base for analyses. When human data to evaluate a functional effect is lacking, animal data is often used.

Step 2: Hazard identification (deficiency and excess end points). From among the factors needed to establish these end points are homeostatic mechanisms, bioavailability, nature of exposure, population variability and age-sex variations.

Step 3: Quantitative evaluation of critical affects. This involves the evaluation of the various dose-response curves for each end-point from deficient and excess exposure.

Step 4: Balanced quantitative assessment to determine acceptable range of oral intake (AROI). The effects of both deficient and excess exposure in healthy populations are considered in the derivation of AROI.

Step 5: Exposure assessment. This identifies and quantifies exposure sources (e.g. water, food, supplements, soil and dust), bioavailability and exposure patterns of the populations.

Step 6: Risk characterization. Integration of the AROI and the exposure information.

CHAPTER 4

MEDICAL GEOLOGY OF FLUORIDE

The link between fluoride geochemistry in water in an area and the incidence of dental and skeletal fluorosis is a well established relationship in medical geology. While the essentiality of fluoride for human health is still being debated, its toxicity has now caused considerable concern in many lands where fluoride is found in excessive quantities in the drinking water. As in the case of some essential trace elements, the optimum range of fluoride varies within a narrow range and this causes fluoride imbalances, very often in large populations, mostly in developing countries of the tropical belt.

In the case of many trace elements, food is the principal source. Much of the fluoride entering the body however is from water and the hydrogeochemistry of fluoride in surface and groundwater is therefore of major interest.

The impact of fluoride on health is shown in Table 4.1. Several organizations throughout the world have considered fluoridating water supplies in areas where the water supplies do not provide the optimum levels of fluoride. This however, has resulted in great disagreement between those who consider fluoride as being beneficial and those who consider it as a poison (Grant, 1986; Sulton, 1988; Colquhon, 1990; Gibson, 1992; Turner et al., 1992; Pendrys, 2001).

Among the dental health concerns, dental fluorosis is the most common manifestation of excessive intake of fluoride-rich water. Children under the age of 7 are particularly vulnerable. Other factors such as nutritional deficiencies, mostly calcium and vitamin C also play minor roles in the disease.

As in the case of iodine deficiency diseases, a large population of the world is affected by fluoride toxicity in drinking water. More than 200 million people including about 70 million in India and 45 million in China (Yang et al., 2003) are prone to fluorosis. The African continent and some parts of South America such as Mexico are also estimated to have millions of people at risk of dental and skeletal fluorosis, though the exact figures are not known.

Table 4.1. Impact of fluoride on health (WHO, 1971)

Concentration of fluoride in drinking water	Impact on health
0.0-0.5 mg/L	Limited growth and fertility, Dental caries
0.5-1.5 mg/L	Promotes dental health resulting in healthy teeth
1.5-4.0 mg/L	Prevents tooth decay
4.0-10.0 mg/L	Dental fluorosis (mottling of teeth)
>10.0 mg/L	Dental fluorosis, Skeletal fluorosis (pain in back and neck bones) Crippling fluorosis

It is worthy of note that unlike Fe in ground and surface water, fluoride does not impart any colour to the water nor does it give any kind of taste to the water. In highly underdeveloped rural areas in the tropics where drinking water is obtained directly from the ground, excessive fluoride in water may turn out to be an invisible poison. People who obtain food and water directly from their home environments for many years are particularly susceptible to diseases caused by excess fluorides in drinking water.

GEOCHEMISTRY OF FLUORIDE

Fluoride is the most electronegative and chemically reactive element of all halides. It is highly reactive with practically all organic and inorganic substances. In the natural environment it occurs as the fluoride ion F^-. It is found mostly in the silicate minerals of the earth's crust at a concentration of about 650 mg/kg (Adriano, 2001).

The geochemistry of the F^- (ionic radius 136 pm) is similar to that of the OH^- ion (ionic radius 140 pm) and there can be easy exchange between them. Extensive research has been carried out on the fluoride-hydroxyl exchange in geological materials (Gillberg, 1964; Stormer and Carmichael, 1971; Ekstrom, 1972; Munoz and Ludington, 1974). Fluorapatite $[Ca_5(PO_4)_3F]$ and hydroxylapatite $[Ca_5(PO_4)_3OH]$ are isomorphic end members in the solid solution series $Ca_5(PO_4)_3(OH,F)$. Hydroxyl apatite, however, is the main mineral phase of enamel in human teeth, a fact often used as an argument against the non-essentiality of fluoride to humans.

As shown by Aswathanarayana et al. (1985), fluorides in the surface and the groundwater are derived from:

(a) Leaching of the rocks rich in fluorine, e.g. granites (750 mg/kg), alkalic rocks (950 mg/kg), volcanic ash and the bentonites (750 mg/kg), phosphatic fertilizers 3.0-3.5%.
(b) Dissolution of fluorides from volcanic gases by percolating groundwaters along faults and joints of great depth and discharging as fresh and mineral springs.
(c) Rainwater which may acquire a small amount of fluoride from marine aerosols and continental dust.
(d) Industrial emissions such as freons, organo-fluorine and dust in cryolite factories.
(e) Industrial effluents.
(f) Run-off from farms using phosphatic fertilizers extensively.

Fluorine is associated with many types of mineral deposits (Boyle, 1974) and hence it is a good indicator of mineral deposits (Lalonde, 1976). The geochemical dispersion haloes of fluorine from mineral deposits are often detected in ground and surface waters, stream sediments and soils. The higher concentrations of fluoride in water and soil are therefore often the result of the occurrence of mineral deposits in the vicinity. The fluorine chemistry of granitic material is relevant to economic prospecting in granitic terrains since fluorine is associated with Sn-W-Mo and REE-Zr-Ta-Be deposits, with Li-Rb-Cs pegmatites, rare-metal greisens and albitized granites and is ultimately responsible for fluorite and cryolite deposits (Bailey, 1977; Correns, 1956).

Fluorine is located in:

(a) F^- rich minerals such fluorite, apatite
(b) Replacement of OH^- and O^{2-} ions in muscovite (mean 0.1-0.3%), biotite (mean ~0.7%), hornblende (mean ~0.2%) and sphene (range 0.1-1.0%)
(c) Solid and fluid inclusions-micas and feldspars, fluid inclusions in quartz
(d) Rock glasses-obsidians and pitchstones.

A list of fluoride-bearing minerals with formulae, fluoride contents and distribution in various granitic materials is given in Table 4.2. It is worthy of note that from among 150 fluoride-bearing minerals listed by Strunz (1970), 63 are silicates, 43 halides and 24 phosphates. In magmatic rocks, only topaz and fluorite contain fluorine as an essential part of the composition.

Geochemistry of Fluoride in Weathering and Solution

As mentioned in Chapter 2, rock and mineral weathering in the tropical climate is intense. Fluoride with its tendency to enter the aqueous medium is therefore leached out from the fluoride-bearing minerals. The geochemical pathways of fluoride in such a physico-chemical environment is strongly influenced by processes involving adsorption-desorption and dissolution-precipitation reactions. It is important therefore to realise that the degree of weathering and the leachable fluoride in a terrain is of greater significance in the fluoride concentration of water than the mere presence of fluoride-bearing minerals in soils and rocks. Some rocks, bearing minerals such as Ca-Mg carbonate act as good sinks for fluoride (Christensen and Dharmagunawardena, 1986). The leachability of fluoride from the carbonate concretions is controlled by (a) pH of the draining solutions (b) alkalinity (c) dissolved CO_2 and the pCO_2 in the soil.

Ramesam and Rajagopalan (1985) studied the fluoride geochemistry in the natural hard rock areas in Peninsular India and summarised a mechanism of fluoride pathways in arid and semi-arid areas as shown in Figure 4.1.

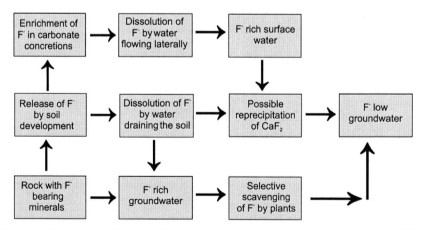

Fig. 4.1. Mechanism for fluoride ingestion in arid and semi-arid areas (modified after Ramesam and Rajagopalan, 1985)

Medical Geology of Fluoride 63

Table 4.2. Fluoride-rich minerals associated with granitic materials (after Bailey 1977)

Name	Formula	F (wt.%)
Fluorite	CaF_2	47.81-48.80
Cryolite	Na_3AlF_6	53.48-54.37
Fluocerite	CeF_3	19.49-28.71
Yttrofluorite	$(Ca,Y)(F,O)_2$	41.64-45.54
Gagarinite	$NaCaYF_6$	33.0-36.0
Bastnasite	$Ce(CO_3)F$	6.23-9.94
Synchisite	$CeCa(CO_3)_2F$	5.04-5.82
Parisite	$Ce_2Ca(CO_3)_3F_2$	5.74-7.47
Pyrochlore	$NaCaNb_2O_5F$	2.63-4.31
Microlite	$(Ca,Na)_2Ta_2O_6(O,OH,F)$	0.58-8.08
Amblygonite	$LiAl(PO_4)$	0.57-11.71
Apatite	$Ca_5(PO_4)_3(F,ClOH)$	1.35-3.77
Herderite	$Ca(BePO_4)(F,OH)$	0.87-11.32
Muscovite	$KAl_2(AlSi_3O_{10})(OH,F)_2$	0.02-2.95
Biotite	$K(Mg,Fe)_3(AlSi_3O_{10})(OH)_2$	0.08-3.5
Lepidolite	$KLi(Fe,Mg)Al(AlSi_4O_{10})(F,OH)$	0.62-9.19
Zinnwaldite	$KLiFe^{2+}Al(AlSi_3O_{10})(F,OH)_2$	1.28-9.15
Polylithionite	$KLi_2Al(Si_4O_{10})(F,OH)_2$	3.00-7.73
Tainiolite	$KLiMg_2(Si_4O_{10})F_2$	5.36-8.56
Holmquistite	$Li_2(Mg,Fe^{2+})_3(Al,Fe^{3+})_2(Si_2O_{22})(OH,F)_2$	0.14-2.55
Hornblende	$NaCa_2(Mg,Fe,Al)_5(Si,Al)_8O_{22}(OH,F)_2$	0.01-2.9
Riebeckite	$Na_2Fe_3^{2+}Fe_2^{3+}(Si_4O_{11})_2(OH,F)_2$	0.30-3.31
Arfvedsonite	$Na_3Fe_4^{2+}Fe^{3+}(Si_4O_{11})_2(OH,F)_2$	2.05-2.95
Ferrohastingsite	$NaCaFe_4^{2+}(Al,Fe^{3+})(Si_6Al_2O_{22})(OH,F)_2$	0.02-1.20
Spodumene	$LiAl(SiO_3)_2$	0.02-0.55
Astrophylite	$(K,Na)_2(Fe^{2+},Mn)_4(TiSi_4O_{14}(OH)_2$	0.70-0.86
Wohlerite	$NaCa_2(Zr,Nb)O(Si_2,O_7)F$	2.80-2.98
Tourmaline	$Na(Mg,Fe)_3Al_6(BO_3)_3(Si_6O_{18})(OH)_4$	0.07-1.27
Sphene	$CaTiSiO_5$	0.28-1.36
Topaz	$Al_2SiO_4(OH,F)_2$	13.01-20.43
Yttrobrithiolite	$(Ce,Y)_3C_2(SiO_4)_3OH$	0.50-1.48

This relationship shows that the concentrations of fluoride are directly proportional to Ca^{2+} concentrations. The absence of calcium in solution allows higher concentration of fluoride to be stable in solution. In areas of alkaline volcanic rocks and in high cation exchange conditions, notably in the presence of clay minerals, such a situation may prevail.

The soil-water and the rock-water interaction and the residence time of fluoride are two important factors in the concentration of fluoride in water. The geochemical environments of surface dug wells and those in the deeper bore holes are different. Whereas the shallow dug wells tend to have lower fluoride concentrations due to leaching and rapid groundwater circulation and lower residence times, the water-rock interaction in the deep bore holes is enhanced by longer residence times. Hence it is quite common to observe higher fluoride concentrations in water obtained from deep wells (Figure 4.2).

Climatic effects, notably evaporation of surface water due to the prevailing high ambient temperature, also affect the relative fluoride concentration in the water. In many arid regions of the world, this situation arises while in regions where there is intense rainfall, leaching of fluoride takes place easily resulting in its lower concentrations.

Fluoride concentrations in natural water vary widely depending on the geochemistry of the rainfall, soils, rocks and minerals in the immediate environment. Recently, Edmunds and Smedley (2004), reviewed the fluoride concentrations in natural water from different geological situations (Table 4.3) and it is observed that some areas such as Lake Magadi in Tanzania have fluoride concentrations as high as 1980 mg/L while in some other regions (e.g. UK chalk regions) it is lower than 0.1 mg/L.

Table 4.3. Average fluoride ranges of different types of waters (summarised from Edmunds and Smedley, 2004)

Source	Range of F^- (mg/L)
Rainfall	0.013 to 0.096
Surface waters (rivers)	0.06 to 0.18
Surface waters in high fluoride regions	0.6 to 1281
Soil water	0.02 to 0.30
Geothermal springs	0.4 to 330
Groundwater crystalline basement rocks	<0.02 to 20
Groundwater volcanic rocks	2.1 to 250
Groundwater sediments and sedimentary basins	<0.1 to 29

Medical Geology of Fluoride 65

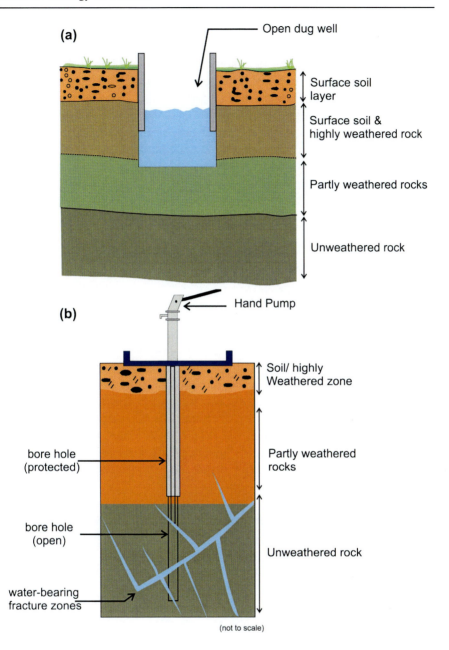

Fig. 4.2. Groundwater regimes in (a) shallow dug well, (b) deep well

Fluoride in Soils

Fluoride in soils is found in the range 20-500 mg/kg (Kabata-Pendias and Pendias, 1984) and its mobility is influenced mainly by pH and complexation with aluminium and calcium (Pickering, 1985). The soil type, salinity and fluoride concentration are also important in the amount of fluoride adsorbed by soil. Acidic conditions favour the fluoride adsorption markedly. Barrow and Ellis (1986) reported that the maximum adsorption of fluoride to soil was at pH 5.5. In acidic soils when the pH is lower than 6, much of the fluoride complexes with either Al or Fe. In the case of alkaline soils at pH >6.5, it is almost completely fixed in soils as calcium fluoride, provided sufficient calcium carbonate is available (Brewer, 1966).

Clay minerals, $Fe(OH)_3$, $Al(OH)_3$ and fine-grained soils adsorb fluoride relatively easily, by displacing hydroxides on the clay surface. It is best adsorbed at a pH range of 3-4 and decreases above 6.5 (Savenko, 2001).

Fluorine is immobile in soil and in soil profiles the fluoride content decreases with increasing distance from the parent rock. Soils therefore act as good sinks for fluoride. Fluoride retention in soil is generally correlated with soil aluminium and it leaches out simultaneously with the leaching of Al, Fe and organic material in the soil (Polomski et al., 1982). In some soils, notably the sandy acidic soils, leaching out of fluoride can be greatly reduced by the addition of lime or gypsum which precipitates fluorite. In the case of calcareous soils with low flow rates of water, the adsorption of fluoride from the water phase is an important step in the geochemical pathway of fluoride in soils. At high flow rates however, this exchange is limited (Flühler et al., 1982).

In agricultural soils where there is a high input of phosphate fertilizer, the fluoride contents may increase significantly phosphate fertilizers (8500-38000 mg/kg) (Kabata-Pendias and Pendias, 1984) and sewage sludges (80-1950 mg/kg) (Rea, 1979), have high fluoride contents.

Fluoride in Sediments

Fluorine is an abundant halogen in sedimentary rocks with fluorite, apatite, mica, illite, and montmorillonite being the main fluorine-bearing minerals (Deshmukh et al., 1995). As shown in Table 4.4, shales, volcano-clastics and bentonites are among the most fluoride-rich sedimentary rocks. Shales in particular, due to their high clay content retain larger concentrations of fluoride. It has been observed that out of the total fluorine content of the clays associated with clastic rocks, 80-90% is found in the mica group of minerals while the rest is adsorbed on clay minerals montmorillonite, illite and kaolinite. The average fluorine content of sedimentary micas and illite is more than 1.5 times the average fluorine content of igneous rocks.

Table 4.4. The average fluorine content in different sedimentary rocks (Fleischer and Robinson, 1963)

Rocks	Range in mg/kg	Average in mg/kg
Limestone	Up to 1210	220
Dolomite	110-400	260
Sandstone and Greywacke	10-1100	200
Shale	10-7600	940
Volcanic ashes and Bentonites	100-2900	750
Oceanic sediments	100-1600	730

Kau et al. (1998) compared the experimental sorption isotherms of fluoride for aluminosilicate clays. It was observed that bentonite is a far superior fluoride sorbent than kaolinite quantitatively even though the mechanism of sorption may be the same. Factors that influenced fluoride sorption included solution pH, clay surface area, aluminium content and the presence of certain exchangeable cations capable of forming fluoride precipitates.

Experiments by Kau et al. (1997) on fluoride retention by kaolin clay showed that the quantity of fluoride sorbed onto kaolin stabilizes essentially after 24 hours. Fluoride measurements after 24 hours revealed that a minimum of 95±2% of the equilibrium concentration has been achieved throughout the tested pH range (Figure 4.3).

Fluoride in Plants

Fluorine is not considered as an essential element for plants. The fluoride levels in terrestrial biota are higher in areas with high fluoride levels from natural and anthropogenic sources. Lichens, for example are used extensively as biomonitors for fluorides. Mean fluoride concentrations of 150-250 mg/kg were measured in lichens growing within 2-3 km of fluoride emission sources, compared with a background level of <1 mg fluoride/kg (Anonymous, 2002). In soil, much of fluoride is insoluble and hence less available to plants. However, since high soil fluoride concentrations or low pH, low clay and/or organic matter contents in soil can increase fluoride levels in solution, there can be greater uptake by the plant. It is of interest to note that due to the low mobility of fluoride within the plant, the root has higher fluoride contents than the shoot.

Terrestrial plants however may accumulate inorganic fluorides from the air through stomata in the leaves. Further, small amounts may also enter the plant through the epidermis and cuticle (Underwood, 1962). The application of fluoride-bearing phosphate fertilizers however introduces a sudden influx of fluoride into the soil and the plants may accumulate more fluoride in the form of aluminium fluoride species (Steven et al., 1997).

Tea (*Camellia sinensis*) has been known to accumulate fluoride in higher concentrations (0.1-0.6 mg/100 ml, Food and Nutrition Board, 1997). It is cultivated in acidic soils in the tropical regions and is the most popular drink next to water. Aluminium accumulates profusely, with Al in tea leaves reaching levels as much as 8700-23000 mg/kg and even up to 30000 mg/kg (Matsumoto et al., 1976). However, as mentioned earlier, the complexation of the fluoride with Al paves the way for fluoride entry into the plant.

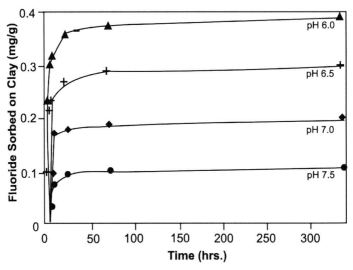

Fig. 4.3. Effect of reaction time on fluoride sorbed onto kaolinite at pH values between 6.0 and 7.5 (Kau et al., 1997)

In areas where fluorosis is endemic, notably in the Asian countries, the fluoride content of food plants grown have shown higher concentrations (Wakode et al., 1993; Kumari et al., 1995). It was observed that the fluoride intake in sorghum, wheat, rice, red gram dhal and red chillies was in proportion to the fluoride concentration of the root zone. The average Partition Factor (F⁻ -food/F⁻ -soil) was of the order in dry red chillies 15.76; red gram dhal 5.3; sorghum 3.6; wheat 2.4 and rice 2.4.

From among aquatic plants, the water hyacinth (*Eichornia crissipes*) is known to take up fluoride from water. Plants that were exposed to fluoride solutions of 6-26 mg/L for 4 weeks had a fluoride uptake of 0.8 mg/kg (Rao et al., 1973).

FLUORIDES AND HEALTH

As mentioned earlier, the range of fluoride tolerance and toxicity is narrow. Deviation from the optimal levels therefore results in dental health effects such as caries and fluorosis. The beneficial and detrimental effects of fluoride are mainly on the tooth enamel and bone. While the prevalence of dental caries is inversely related to the concentration of fluoride in drinking water, the prevalence of dental fluorosis is strongly associated with the concentration of fluoride, with a positive dose-response relationship. In the case of skeletal fluorosis, apart from a high intake of fluoride-rich water, other factors such as nutritional status and diet, climate (related to fluid intake) concomitant exposure to other substances and intake of fluoride from sources other than drinking water are believed to play an important role in the development of this disease (WHO, 2002). Other diseases such as cancer, respiratory, hepatic, renal and haematopoietic disorders attributed to fluorides have not been proved beyond doubt.

Bioavailability of Fluoride

Figure 4.4 illustrates the fate of fluoride ingested with food. In areas where the natural fluoride levels are low the total daily adult intake will be <1 mg while for those living in areas with a higher water fluoride intake, the total daily fluoride intake will be about 2 mg/day.

As shown in Figure 4.4 the fluoride ingested is absorbed by the stomach and in greater quantities by the small intestine. In the stomach, when the pH is low, highly diffusible hydrogen fluoride forms (pKa = 3.4) and this results in high absorption (Whitford and Pashley, 1984). In the small intestine fluoride is absorbed as the free ion (F⁻) by diffusion via membrane channels and is non-pH dependent (Nopakun and Messer, 1990). When fluoride is ingested with little or no food, fluoride absorption is much higher in the stomach, some times as much as 100% (Ophang, 1990).

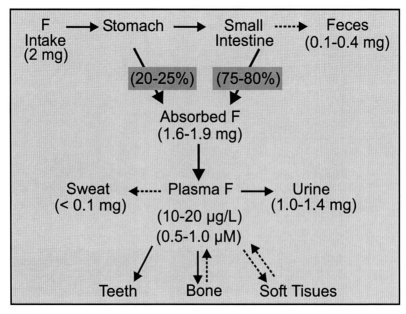

Fig. 4.4. Fate of fluoride ingested with food in adult. Dashed arrows denote a minor pathway (Cerklewski, 1997a)

Under normal fluoride intakes and rate of absorption, the plasma fluoride concentration is known to range from 10-20 µg/L or 0.5-1.0 µM (Ekstrand, 1978). In the adult body over 95% of total fluoride is in the bones and teeth and these remove the plasma fluoride rapidly by ion exchange with hydroxyl ions, citrate and carbonate. Soft tissues on the other hand do not accumulate fluoride. Forbes (1990) has noted that the total body fluoride estimated to be about 2.6 g is second only to the trace element iron. It is during the period of rapid development of bones and teeth that fluoride is taken up quickly. As in the case of bone, the outer layer of surface enamel contains higher fluoride- about 3000 µg/g as against 100 µg/g at the deeper dentine-enamel junction (Ten Cate and Featherstone, 1996).

Fluoride homeostasis in the body is maintained by the combined effects of fluoride assimilation by bone and urinary excretion bearing in mind that soft tissues do not take up fluoride. When the fluoride uptake by bone is low, urinary excretion of fluoride increases.

Other elements also influence the fluoride homeostasis by interaction and complex formation mostly in the alkaline small intestine. Calcium and magnesium form insoluble complexes with fluoride thereby decreasing fluoride uptake by bone and teeth (Cerklewski, 1997b). However, if the fluoride is in the form of monofluorophosphate in contrast to sodium fluoride, there is no effect by the presence of calcium (Villa et al., 1992).

The interaction of aluminium with fluoride has been the subject of several studies (Lubkowska et al., 2002; Ahn et al., 1995). Aluminium is known to suppress the uptake of fluoride by modifying the metabolism of phosphorus, calcium, magnesium and fluorine and enhances the development of osteomalacia and osteodystrophia (Ahn et al., 1995).

Dental Fluorosis

Figure 4.5 illustrates the cross section of a human tooth. The outermost layer of the exposed tooth is a hard thin, translucent layer that envelopes and protects dentin, the main portion of the tooth structure. Enamel, dentin and cementum are all composite materials composed of the mineral hydroxylapatite (HA) (Figure 4.6), protein and water. Enamel is the hardest substance found in the human body and has ~90% mineral, 1% organic matter and 3% water. On account of its hardness, it is also brittle. Enamel has the carbonate- rich apatite arranged in enamel rods or prisms 4-5 µm in diameter (Marshall et al., 2003). These are held together by a cementing substance and surrounded by an enamel sheath. The rods that make up enamel are formed by cells known as ameloblasts.

The mineral apatite $Ca_5(PO_4)_3(OH, F, Cl)$ has varying amounts of fluorine, chlorine or the hydroxyl group, though in some cases one of these may approach the 100% level, and according to the major presence of F^-, Cl^- or OH^-, they are termed fluorapatite, chloroapatite and hydroxyapatite, respectively.

When fluoride is present in the water ingested, some of it is incorporated into the apatite crystal lattice of the tooth enamel during its formative stages, the enamel becomes harder and discolouration results (Figure 4.7).

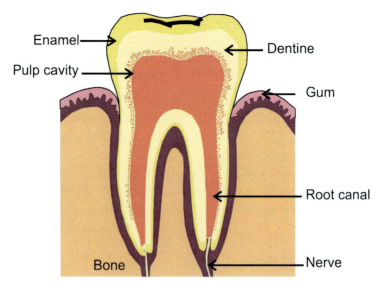

Fig. 4.5. Schematic cross section of a human tooth

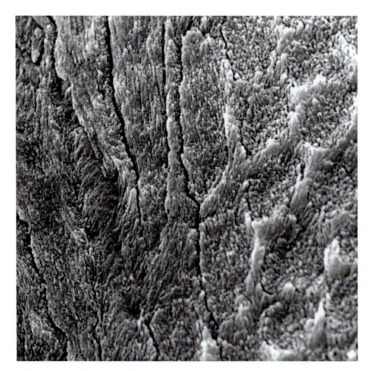

Fig. 4.6. Scanning Electron Microscope image of the dental enamel of a front tooth with well orientated rod structure from apatite (photo courtesy of Prof. Minoru Wakita, Hokkaido University, Japan)

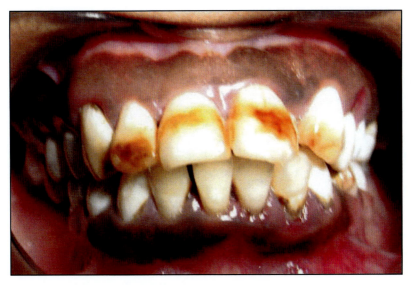

Fig. 4.7. Discolouration of teeth in dental fluorosis

Clinically the appearance of enamel fluorosis is known to vary with the amount of fluoride ingested during early childhood, appearing in its mildest forms as a white flecking of the enamel coalescing to become more visible in its moderate forms and marked by a dark brown staining of the enamel, with actual breakdown of the enamel in the most severe cases (Dean and McKay, 1939, Rozier, 1994). Table 4.5 shows the classification of dental fluorosis. Even though the WHO guidelines have recommended an upper limit of fluoride concentration in drinking water as 1.5 mg/L, in many tropical countries where there is a high sweat loss and a high intake of water due to the hot weather such an upper limit may be unsuitable (Brouwer et al., 1988).

In several other tropical countries (India-Handa, 1975; Tanzania- Aswathanaryana et al., 1985; Kenya- Manji et al., 1986; Sri Lanka- Warnakulasuriya et al., 1992; Ghana- Apambire et al., 1997) it has been found that the WHO recommended levels for fluoride in drinking water are not acceptable and that the incidence of dental fluorosis is common even in areas with lower levels of fluoride in water. Brouwer et al. (1988), who observed that the WHO guidelines were unsuitable for Senegal, showed that in the hot climate of Senegal, both dental and skeletal fluorosis are more prevalent and more severe than would be expected from the WHO recommended maximum limits. They found that 66.5% of children had mild dental fluorosis at the 1.0 mg/L level and beyond the 4.0 mg/L level dental fluorosis had reached 100%.

Table 4.5. Dental fluorosis classification by Dean (1942)

Classification	Criteria-Description of Enamel
Normal	Smooth, glossy, pale creamy-white translucent surface
Questionable	A few white flecks or white spots
Very Mild	Small opaque, paper-white areas covering less than 25% of the tooth surface
Mild	Opaque white areas covering less than 50% of the tooth surface
Moderate	All tooth surfaces affected; marked wear on biting surfaces; brown stain may be present
Severe	All tooth surfaces affected; discrete or confluent pitting; brown stain present

In Sri Lanka, Warnakulasuriya et al. (1992) showed that the optimal level of fluoride in groundwater for caries protection should be 0.6-0.9 mg/L. Even at these moderate levels, only 34% of the children were entirely fluorosis free. They recommended that the WHO limit of 1.5 mg/L fluoride in drinking water is not appropriate for hot and dry climates and that an upper limit of 0.8 mg/L is more suitable.

Figure 4.8 illustrates the Community Fluorosis Index (CFI) where

$$CFI = \sum \frac{(\text{number of children} \times \text{Dean's Index Score})}{(\text{Total number of children examined})}$$

and the Dean's Index Score classifies individuals into 5 categories depending on enamel alteration. Figure 4.9 shows the relationship between fluoride concentration of drinking water and dental caries/dental fluorosis. It can be observed that the CFI value detrimental to human dental health varies with mean annual temperature (Minoguchi, 1974).

Accordingly if the CFI is below 0.4, the water is considered safe from a health point of view. If the value is above 0.6, the excess quantity of fluoride in drinking water in the area has to be removed. CFI in the range 0.6 to 3.5 often indicate the onset of dental fluorosis and a CFI greater than 3.5 points to the possibility of skeletal fluorosis.

Medical Geology of Fluoride 75

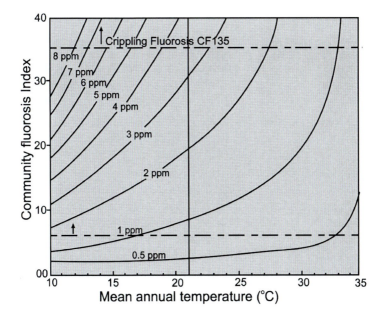

Fig. 4.8. Relationship between community fluorosis index and the mean annual temperature (Minoguchi, 1974)

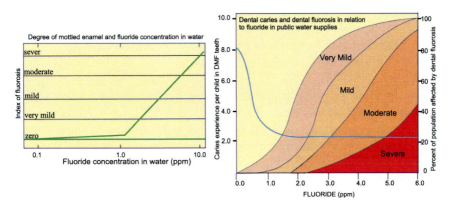

Fig. 4.9. Degree of enamel mottling in relation to fluoride concentration in water (Latham et al., 1972)

Skeletal Fluorosis

As discussed above, high ingestion of fluoride-rich water causes dental fluorosis. If the levels of fluoride ingestion are still higher, and if it continues over several years, a serious debilitating disease which affects the bones, termed skeletal fluorosis (Fig. 4.10) results. This may even cause neurological complications. About 96-99% fluoride retained in the body combines with mineralised bones and when the ingestion is >4 mg/day, 50% is retained by the skeleton and the rest is excreted through urine. The skeletal fluoride concentration is known to increase almost proportionately to the amount of fluoride ingested and its resident time (Spencer et al., 1975). There are several varieties of bones in the skeleton and the fluoride content of these vary with pelvis and vertebrae having higher fluoride contents than limb bones. Once incorporated into the hard tissues, the fluoride is retrievable though with difficulty and involves a very slow process of osteoclastic resorption spread over many years. The progress of skeletal fluorosis symptoms are as follows (Reddy et al., 1969):

(a) Vague discomfort and paraestheria in limbs and trunk.
(b) Pain and stiffness in back.
(c) Stiffness increases steadily with restriction of movement.
(d) Appearance of a "poker- back" spine.
(e) Spread of stiffness to various joints.
(f) Fluorosis deformity at hips, knee and other joints.
(g) Bony exostosis seen in limb bones.

In the endemic regions, crippling skeletal fluorosis occurs between the ages of 30 to 50 years. According to Siddiqui (1955), newcomers to a hyperendemic region may develop symptoms within years of their arrival. The main factors which influence the development of skeletal fluorosis (Reddy, 1985) are (a) high fluoride intake (b) continual exposure to fluoride (c) strenuous manual labour (d) poor nutrition (e) impaired renal function due to the disease.

Contrary to earlier thinking, if the levels of fluoride in the drinking water of a region are extremely high, younger age groups including children are also affected. It should be borne in mind that the incidence of fluorosis is higher in tropical countries on account of the higher quantity of water (fluoride-bearing) consumed. Poor nutrient intake may also be a significant factor.

Medical Geology of Fluoride 77

Fig. 4.10. A case of skeletal fluorosis due to ingestion of excessive fluoride from drinking water (photo: courtesy of Ministry of Water, The United Republic of Tanzania)

In its simplest terms the mechanism of the onset of skeletal fluorosis is that in order to immobilise fluoride from circulating fluids in the body, excess fluoride is fixed in the hydroxyl apatite material of the bone by replacement of OH^- by F^- irreversibly till the exposure continues. During this process, the rate of synthesis of bone material (hydroxyl apatite) is considerably increased and this leads to bone formation or osteosclerosis, as seen in those with skeletal fluorosis (Teotia and Teotia, 1992). With the deposition of calcium fluoro-apatite, the bone density and bone mass increases. In the backbone, the perforations through which nervous and blood vessels pass through, are constricted and this leads to pressure on nerves and blood vessels resulting in paralysis and extreme pain.

Dissanayake et al. (1994) have reported a case of skeletal fluorosis with spinal cord compression from Sri Lanka, following consumption of water with high fluoride content for about 20 years (Figure 4.11). The fact that fluoride is not entirely irreversibly bound to the bone has been demonstrated in persons who had lived in areas of fluoride- rich drinking water and subsequently moved to fluoride-low areas. The urinary fluoride concentration in these individuals decreased gradually over long periods indicating that fluoride was being mobilized continuously from the skeleton and subsequently excreted (Hodge et al., 1970).

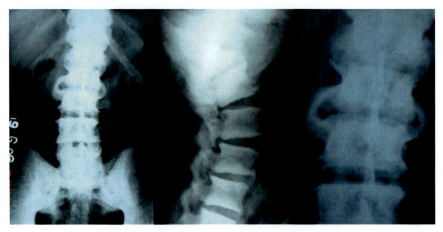

Fig. 4.11. X-ray pictures taken from a 33 year old male skeletal fluorosis patient from Sri Lanka, showing generalized increase of bone density with cortical thickening and coarse tubercular pattern. Ligamentous calcification in the spine and pelvis is observed. Bridging osteophytes in the spine are clearly seen in the picture (photo courtesy Dr. Tilak Abeysekara)

CASE STUDIES

Dental Fluorosis in Sri Lanka

Sri Lanka, a developing country with a population of nearly 20 million and lying in the tropical belt, has well defined wet and dry zones. In the dry zone, dental fluorosis is highly prevalent and a population of about one million people is affected. In some areas in Sri Lanka, dental fluorosis has been recorded as high as 80-98% (Warnakulasuriya et al., 1992; Nunn et al., 1994).

The close relationship between the physical environment and community health is very clearly seen in Sri Lanka. This is mainly due to the large majority of the population of Sri Lanka living in close association with the actual physical environment with only about 25-30% of the population having piped water. Further, nine out of the ten great soil groups are present in Sri Lanka and the effect of the chemistry of the soil and water on the health of the population is seen quite prominently in Sri Lanka (Dissanayake, 1984b; Dissanayake and Weerasooriya, 1987).

Geologically Sri Lanka consists of over 90% metamorphic rocks of presumed Precambrian age and these form 3 major units, the Highland complex, Wanni complex and the Vijayan complex (Figure 4.12). A suite of metasedimentary and metavolcanic rocks formed under granulite facies conditions comprises the Highland complex. Among the metasediments, quartzites, marbles, quartzo-feldspathic gneisses and meta-pelites form the major constituents. In the south-western part, calciphyres, charnockites and cordierite-bearing gneisses are present.

The Wanni complex consists of leucocratic biotite gneisses, migmatites, pink granitic gneisses and granitoids with compositions varying from granitic, syenitic to granodioritic. The granitoids frequently have enclaves of amphibolite and hornblende gneiss. The Vijayan complex is composed of biotite-hornblende gneisses, granitic gneisses and scattered bands of meta-sediments and charnockitic gneisses. Small plutons of granites and charnockites also occur close to the east coast (Cooray, 1978).

It is apparent from the lithology that there are abundant fluoride-bearing minerals such as micas, hornblende, sphene and apatite. Further, minerals such as fluorite, tourmaline and topaz are also found in many locations and these also contribute to the general geochemical cycle of fluorine in the physical environment.

A further source of fluorine is probably the lower crustal volatiles of which fluorine is known to be an important component. These volatiles are found in deep-seated fractures and lineaments and hence tend to get concentrated in materials associated with such faults and fractures.

Vitanage (1989) studied the Precambrian and later tectonic events in Sri Lanka and observed that the rocks of all three complexes are dissected by about 4330 lineaments with lengths varying from 1.25 km to over 100 km in an area of 30,000 km^2 investigated. It is conceivable that many of these lineaments are deep seated and hence the loci for fluorine outgassing.

These could well be of great significance in the general geochemistry of fluorine in Sri Lanka.

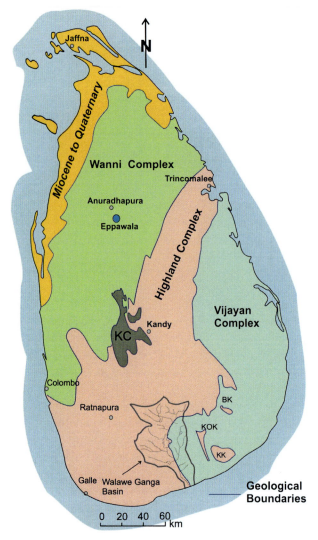

Fig. 4.12. Simplified geological map of Sri Lanka (KC- Kadugannawa Complex; KK- Kataragama Complex; BK- Buttala Klippe; KOK- Kudaoya Klippe)

The plate tectonic model for the geological evolution in Sri Lanka as put forward by Munasinghe and Dissanayake (1982), envisages the central Highland Complex of Sri Lanka to have been a highly metamorphosed assemblage of marine deposited volcanic, volcano-clastic and sedimentary rocks, which evidently carried significant quantities of chlorine and fluorine concentrated in the marine-based sediments (Wedepohl, 1972). The later metamorphism and repeated deformation of these volatile-rich materials within the central Highland Basin probably incorporated fluorine in the minerals leading to an enrichment of fluorine in the crustal rocks of Sri Lanka.

Distribution of fluoride in the groundwater of Sri Lanka

Figure 4.13 illustrates the distribution of fluoride-rich groundwater in Sri Lanka. In the compilation of the Hydrogeochemical Atlas of Sri Lanka, Dissanayake and Weerasooriya (1986) delineated the fluoride zones of Sri Lanka based on the fluoride content in dug well water samples. The histograms for fluoride concentrations in groundwater in seven fluoride-rich districts of Sri Lanka are shown in Figure 4.14 (Raghava Rao et al., 1987; Christensen and Dharmagunawardhena, 1987). The high fluoride areas coincide with the high dental fluorosis areas of Sri Lanka.

Data obtained from a large number of deep wells indicate that a large part of the landmass of Sri Lanka is fluoride-rich. Several regions, in the North Central Province, notably the Anuradhapura and Polonnaruwa districts have fluoride concentrations in groundwater often reaching 10 mg/L and these areas, as expected, have a very high incidence of dental fluorosis.

It is of interest to correlate the fluoride-high and fluoride-low areas delineated with natural factors such as climate and geology. Low-fluoride areas are situated mainly in the wet zone where the average annual rainfall exceeds 5000 mm in certain instances. In these regions, leaching of soluble ions is high and fluoride is rapidly leached. In the dry zone, however, evaporation is high and this brings the soluble ions upwards due to the capillary action in soils. This, although not the sole explanation for the observed distribution of fluoride in the groundwater of Sri Lanka, could nevertheless be a major factor (Dissanayake, 1991a).

In the study of geochemical distribution of fluoride in the groundwater of Sri Lanka, the geology of the terrains needs special consideration. The composition of the rocks in the area, particularly the easily leached constituents coupled with the climate is the key factor in the geochemical distribution of elements in a tropical region. The abundance of fluoride in the rocks and the ease with which it is leached under the effect of groundwater has an important bearing on the abundance of fluoride in the groundwater of the areas concerned and hence the prevalence of dental diseases.

Raghava Rao et al. (1987) studied the fluoride concentration in the groundwater as influenced by the lithology of the terrain (Fig. 4.14). His observations are as follows.

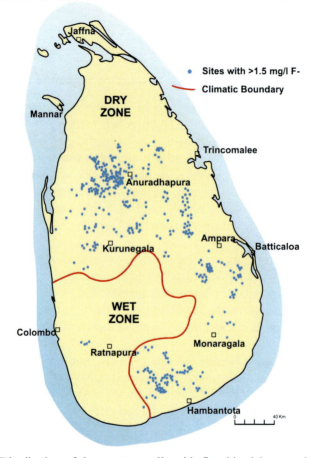

Fig. 4.13. Distribution of deep water wells with fluoride-rich ground water in Sri Lanka (modified after Dissanayake, 1991a)

Anuradhapura district - High fluoride levels throughout the district. Associated rocks: charnockites, charnockitic gneisses, granitic gneisses and hornblende gneisses (granitised through pegmatite intrusions).

Kurunegala district - High fluoride areas (3-5 mg/L), associated with charnockites, intrusive granites, hornblende biotite gneisses and granitic gneisses. Moderately rich fluoride areas (2-3 mg/L) were recognized in garnet-biotite-sillimanite gneisses associated with intrusive granites and granitic gneisses.

Ratnapura district - Fluoride-rich areas were recognised in the undifferentiated metasediments (granitic in composition) and charnockitic gneisses associated with pegmatites and intrusive granites. Marbles and calc-gneisses also constituted a rock assemblage associated with high fluoride zones.

Ampara district - High fluoride areas associated with hornblende-biotite gneisses and granite gneisses/augen gneisses.

Monaragala district - A cluster of high fluoride wells were associated with charnockites and hornblende gneisses.

Matale and Polonnaruwa districts - Christensen and Dharmagunawardana (1987) observed that in the Matale and Polonnaruwa districts high fluoride concentrations are associated with charnockites, biotite gneisses, granulites and gneisses.

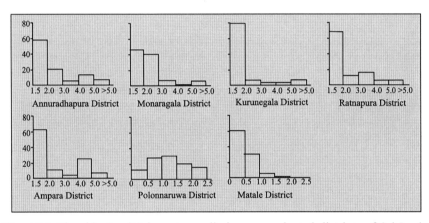

Fig. 4.14. Fluoride (mg/L) in water wells in some selected districts of Sri Lanka (Raghava Rao et al., 1987)

A further point of interest in the geology of Sri Lanka which has a bearing on fluoride geochemistry is the presence of a mineralized belt at the boundary of the Highland-Vijayan complexes (Munasinghe and Dissanayake, 1982). Fluorine, being a volatile element is known to be abundant in such tectonic zones and is enriched in rocks at such locations. Granites are generally rich in fluorine and such granites are found in abundance in the Vijayan Complex. Mineralogically, 30-90% of the fluorine in calc-alkaline granites is generally located in biotite with lesser amounts in hornblende, muscovite, quartz, and accessories. However, accessory minerals-apatite, sphene, fluorite, microlite, pyrochlore, topaz, tourmaline, spodumene, cryolite among others, occasionally contribute more than 50% of the F notably in F-rich magmatic and metasomatic roof-zone granites (Bailey, 1977). Dissanayake and Weerasooriya (1986) observed high fluoride concentrations along the mineralised boundary.

Nanayakkara et al. (1999) studied the prevalence and severity of dental fluorosis in a high fluoride area at Eppawala located near a fluoride-bearing (1.5-2.4%) large exploitable apatite deposit in the North Central province of Sri Lanka. The fluoride levels in the drinking water ranged from 0.21 to 9.8 mg/L, and 97% of the children in the area were affected by dental fluorosis. About 20% had severe fluorosis (scores of 3 and 7). The prevalence of caries increased as the degree of fluorosis increased. The fluoride content in water of deep wells was much greater than in that of surface well samples and this was a major cause of the dental fluorosis in the area.

A study on the dental fluorosis in the Walawe Ganga basin in the south of Sri Lanka by Van der Hoek et al. (2003) showed that prevalence of dental fluorosis among 14 yr old students of the area was 43.2%. In this area, too, drinking water obtained from surface water sources had lower fluoride levels (median 0.22 mg/L) than water from deep tube wells (median 0.80 mg/L). The study by Warnakulasuriya et al. (1992) on 380 children of about 14 years, living in 4 geographic areas of Sri Lanka with fluoride levels of 0.09 mg/L to 8.0 mg/L showed that even in low-fluoride areas dental fluorosis is still prevalent. Their studies were comparable to the findings from other tropical countries such as Senegal and Kenya.

Dental Fluorosis in India

It has been estimated that about 62 million people in 17 out of the 32 states in India are affected by dental and/or skeletal fluorosis, the extent of fluoride contamination of water varying from 1.0 to 48.0 mg/L (Susheela, 1998). Table 4.6 and Figure 4.15 show the districts and states in India affected by dental fluorosis. Rajastan is one of the worst affected regions in India where all 32 districts are affected. With a population of over one billion, provision of safe drinking water in India is a task of gigantic proportions, bearing in mind that around 300 million people still live in absolute poverty in both rural and urban areas. About 6 million children in India are estimated to be affected by high fluoride ingestion (Sharma, 2003).

Table 4.6. Grading of states based on the extent of endemicity (after Susheela, 2003)

Name of the State	Total Districts	No. of Effected Districts
Andhra Pradesh	23	16
Gujarat	19	18
Rajasthan	32	32
Karnataka	27	18
Orissa	32	18
Punjab	17	14
Maharashtra	32	10
Madhya Pradesh (MP)*	45	16
Haryana	19	12
Bihar*	41	6
Tamil Nadu	29	8
Uttar Pradesh (UP)*		18
West Bengal	18	4
Kerala	14	3
Assam	23	2
NCT of Delhi	13	4
Jammu and Kashmir	14	1

* Undivided states of MP, UP and Bihar.
"NCT" stands for National Capital Territory (Note there are no districts rather it is divided into 13 Blocks.)

Choubisa (2001) who investigated endemic fluorosis in 21 villages in Southern Rajasthan where the fluoride concentrations ranged from 1.5 to 4.0 mg/L, observed a maximum prevalence of dental fluorosis (77.1%) in the 17-22 age group. In the other parts of Rajasthan, there are reports of fluoride concentrations in the drinking water as high as 18.0 mg/L (Somapura, 1998). In some areas of Rajasthan where the fluoride levels in the water are very high, intrusions of granite and rhyolites in metasediments have been observed. Mineralization, notably fluorite are also present (Maithani et al., 1998). High fluoride zones correlated with granites, acidic volcanic rocks and dykes. The water-rock interaction was prolonged due to longer residence times and this resulted in higher fluoride concentrations.

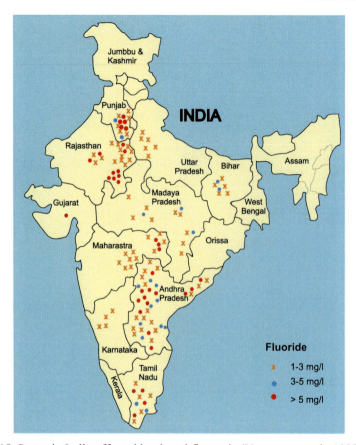

Fig. 4.15. States in India affected by dental fluorosis (Vannappa et al., 1999)

In an interesting case study Saha and Sharma (2002) noted that in the state of Bihar, fluorosis was increasing with time. Earlier work (Ghosh et al., 1986) had shown that in the alluvial plain of Bihar, fluorosis was very low (Table 4.6). Later workers after several years reported high fluoride concentrations from different parts of the state of Bihar (Sharma, 2001), underlain by granite gneisses complexes. Areas underlain by alluvium had also excessive fluoride in groundwater.

The work of Saha and Sharma (2002) showed that over a period of 10 years, the average fluoride content had increased from 0.56 mg/L to 1.3 mg/L with a greater increase recorded in the marginal alluvium than in areas underlain by hard rocks. The correlation between Ca and F which was -0.084 in 1992 had increased to -0.375 in 2002. The fluoride-bearing minerals of the mica schists and granitic gneisses were considered to be the sources of the fluoride. Further, a comparison of the water level distribution with fluoride distribution in 2002 showed that fluoride accumulates more where there is a shallow water table and where seasonal fluctuations are minimal. These areas were underlain by 30 to 140 m thick unconsolidated sediments. There was also $CaCO_3$ precipitation which caused the fluoride content to increase on account of the prevalent pH range of 7.3 to 7.8.

In the state of Karnataka, fluorosis has been reported in several areas, notably in the hard rock terrains. In all the districts, groundwater is the main source of drinking water. The fluoride concentrations in the drinking water ranged from 1.5 to 18 mg/L with about 90% of school children affected by fluorosis. The geology plays a major role as shown by the higher fluoride concentrations in areas underlain by gneisses, schists and quartzites. Some formations had fluoride-rich mafic minerals that yielded higher fluoride concentrations in the associated groundwater (Vannappa et al., 1999). In the Hasan district, the amphibolites, apatite-rich granitic intrusions and dolerite dykes were considered to be the source materials for the fluoride.

Fluorosis in the East African Rift Valley

The Great Rift Valley (Figure 4.16) is a major geological feature in Africa caused by the separation of the African and Arabian plates about 35 million years ago. It extends for over 5000 km from northern Syria to Mozambique and has a width varying from 30-100 km and a depth ranging from a few hundred to several thousand metres. In the eastern part of Africa, the valley is split into 2 parts, the Eastern Rift and the Western Rift. The Western Rift has some of the highest mountains in Africa and deep lakes such as the 1470 metre deep Tanganyika Lake and Lake Victoria.

In the East African Rift Valley, some of the terrains such as those in the north of Nairobi in Kenya are lying at low levels. The lakes in these regions are characterized by high mineral contents and marked salt formation by strong evaporation. This is exemplified by Lake Magadia Soda Lake, and Lakes Elmenteita, Baringo, Bogoria and Nakuru all of which are strongly alkaline.

Due to the weakening of the crust caused by the geological processes of rift formation there are several volcanic mountains. Among these are Mount Kilimanjaro, Mount Kenya, Mount Karisimbi and the crater Highlands in Tanzania.

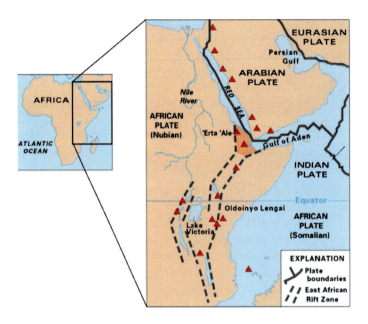

Fig. 4.16. East African rift valley (source: USGS)

These volcanic activities bring about large amounts of volatile releases of CO_2, SO_2, H_2S, HCl and fluorine in the form of HF. In several countries in the East African Rift Valley, fluorosis among people is very high and these volcanic emanations are a primary source of fluoride.

In a recent study on the Ruapehu volcano, New Zealand, Cronin et al. (2003), highlighted the environmental and health hazards of fluoride in volcanic ash. The vent-hosted hydrothermal system of Ruapehu volcano is normally covered by about 10 million m^3 acidic Crater Lake water accumulating volcanic gases. These authors observed that the total F in the ash is often enriched by a factor of 6 relative to original magmatic contents. The carriers of F were considered to be the low soluble minerals such as CaF_2, AlF_3 and $Ca_5(PO_4)_3(OH,F)$. The fluoride therefore was released over long periods of time. In the case of Ruapehu volcanic emissions, several thousand sheep died due to acute fluorosis.

In Kenya the water resources are very limited, particularly in the drier northern regions where groundwater is the main source of drinking water. Table 4.7 shows the percentage distribution of high fluoride groundwaters in Kenya. It is observed that certain provinces such as Nairobi, Rift Valley; Eastern and Central Kenya have highly fluoride-rich groundwater. Clarke et al. (1990) defined the unmodified waters in the Rift Valley as waters whose chemical composition is derived from normal water-rock interactions at moderate temperatures. Very high fluoride concentrations up to 180 mg/L had been observed in these waters, indicative of highly significant leaching of fluoride-rich Rift Valley volcanic rocks. Gaciri and Davies (1993) observed that the groundwaters associated with these volcanic rocks have high alkalinity and were high in Na, K, HCO^-_3, Cl^-. Ca and Mg were found in low concentrations due to their precipitation as carbonates. It had been observed that the highest levels of fluoride in groundwaters in the Rift Valley were 39.0 mg/L from wells and 43.5 mg/L in bore holes (Wilkister et al., 2002).

Table 4.7. Percentage distribution of high fluoride groundwaters in Kenya (Nair et al., 1984)

Province	0.1-0.4	0.5-1.0	1.1-3.0	3.1-5.0	5.1-8.0	>8.0	Number of Samples
Nairobi	9.8	9.8	19.7	13.2	15.8	31.7	183
Rift Valley	14	15.7	38.7	13.7	8	9.9	313
Eastern	11.7	23.1	37	6.1	9.4	12.7	181
Northeastern	9.3	28.9	44.7	9.2	3.9	3.9	76
Central	25.8	21.2	30.3	9.1	5.3	8.3	396
Nyanza	25.8	29.1	25.8	9.7	6.4	3.2	31
Coast	40.9	22.6	26.8	5.4	1.1	3.2	93
Western	77	15	-	-	8	-	13
							1286
Total for Kenya	19.3	19.2	31.9	10.1	7.7	11.8	100%

Apambire et al. (1997) studied the geochemistry, genesis and health implications of fluoriferous groundwaters in the upper regions of Ghana and observed that groundwater fluoride levels reached 0.11 to 4.60 mg/L with a mean value of 0.97 mg/L. The highest concentrations were associated with hornblende granite and syenite rock suites (Figure 4.17). The areas of anomalous groundwater fluoride coincide largely with terrains underlain by rocks of the Bongo Granitic Suite which comprised of coarse-grained hornblende-granite grading towards its outer contacts into biotite and hornblende syenitic phases. The highest groundwater fluoride concentrations were found in the interior phases of the intrusions.

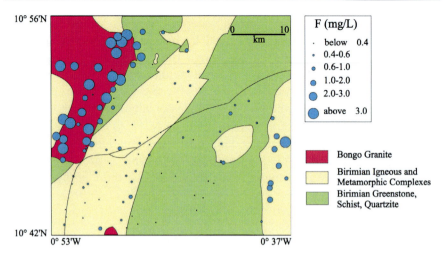

Fig. 4.17. Distribution of fluoride in groundwater in the Bolgatanga area of Northern Ghana. The highest concentrations are associated with granite outcrops and coincide with the incidence of dental fluorosis (Smedley et al., 1995; reproduced with kind permission from the British Geological Survey)

The anomalous fluoride contents in the groundwater associated with the Bongo granitoids were thought to have arisen from fluoride present and to a lesser extent from both dissolution and anion exchange from micaceous minerals and their altered clay products.

Apart from the geology, the climate, as in the case of Sri Lanka, also plays a major role in the enrichment of fluoride in the groundwater of Ghana. The arid zones of the country are notably high in fluoride-rich water formations. Dental fluorosis is also common in such areas. The groundwaters in granitic rocks of the south-west plateau had lower fluoride concentrations due to high rainfall and dilution (Smedley et al., 1995). The fluoride concentrations in parts of northern Tanzania are probably the highest in the world. Aswathanarayana et al. (1985) have recorded abnormally high fluoride contents in the Maji Ya Chan River (21-14 mg/L), Engare Nanyuki river (21-26 mg/L), pond waters of Kitefu (61-65 mg/L), thermal springs of Jekukumia (63 mg/L) and in the soda lakes of Momella (up to 690 mg/L).

The geochemical model proposed by Aswathanarayana et al. (1985) (Figure 4.18) accounts for the enhanced fluoride level as being due to episodic influx of fluoride from volcanic ash exhalations and sublimates related to Miocene to Recent Volcanism. The alkaline volcano of Oldoinyo Lengai had erupted in 1960. Since such geological sources enrich the groundwater in fluoride, there are severe cases of dental and skeletal fluorosis in many parts of Tanzania.

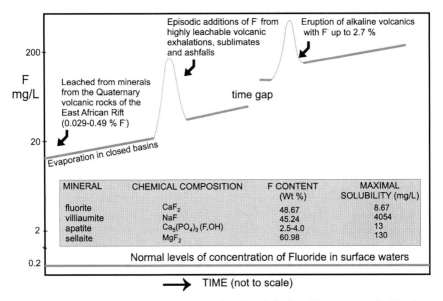

Fig. 4.18. Conceptual model to account for the high fluoride contents in Northern Tanzania (modified after Aswathanarayana et. al., 1985)

In the Rift Valley of Ethiopia, some lakes such as Abiyata and Shalla, located on the rift floor have fluoride concentrations reaching levels as high as 300 mg/L (Chernet and Travi, 1993). These water characteristics have rendered the water rather unsuitable for agriculture. Chernet et al. (2001) subjected chemical analyses of 320 samples taken throughout the region for geostatistical, chemical equilibria and simulation of evaporation-concentration processes using the computer software AQUA.

Their results showed that the water associated with the volcanic rocks have positive alkalinity residual of calcite. Due to the strong evaporation in the arid to semi-arid conditions, calcite precipitation causes a decrease in the chemical activity of calcium, resulting in an increase of fluoride concentrations controlled by equilibrium with CaF_2. When the pH reaches values as high as 9 to 10, fluoride accumulates in the lower zones of the basins. Chernet et al. (2001) showed that the increase in fluoride content and the alkaline-sodic characteristics depend mainly on the unbalanced initial stage between the carbonate alkalinity and calcium. $[(HCO^-_3)>2(Ca+Mg)]$, resulting from the weathering and solution effects of volcanic rocks.

In a more recent study Reimann et al. (2003), collected drinking water samples throughout the Ethiopian part of the Rift Valley and analysed them for 70 chemical parameters. It was observed that fluoride was the most problematic element with 33% of the samples having values over 1.5 mg/L and up to 11.6 mg/L. The incidence of dental and skeletal fluorosis, as expected, was very high.

Endemic Fluorosis in China

In terms of both incidence and severity, China is one of the countries most seriously affected by endemic fluorosis. It occurs in more than 30 provinces, municipalities and autonomous regions affecting a population of 45 million (Ministry of Health PRC, 1997). Figure 4.19 illustrates a map showing the distribution of fluorosis districts in China. Endemic fluorosis areas in China are divided into 6 types according to the fluoride source (Wang et al., 2002);

    (a) shallow groundwater of high fluoride contents (>1.0 mg/L)
    (b) deep groundwater with high fluoride
    (c) hot springs with high fluoride
    (d) abundant fluoride - bearing rock formations
    (e) high fluoride coal
    (f) high fluoride tea

In the large semi-arid and arid regions of China, the enrichment of fluoride in groundwater is due to the leaching of fluoride from the fluoride-bearing rock formations and enrichments in water due to rapid evaporation, aided by poor drainage. In these areas the fluoride concentration is around 5 mg/L.

Zheng and Hong (1988) recorded maximum fluoride contents in surface and phreatic waters as 129 mg/L and 40 mg/L, respectively. In the Mt Da Xinganling and Mt Yanshan areas there are fluoride-bearing volcanic and intrusive rocks (500-800 mg/kg) and these are probable sources of the fluoride. Further, loess present in these regions also acts as carriers of fluoride, the F^- in loess ranging from 490 to 550 mg/kg. Soda salinisation of soils in the affected regions is also considered to be an important factor in the fluoride enrichment.

There are 2493 recorded hot springs in China (Zheng and Hong, 1988). These are distributed in Yunnan, Guangdong, Fujian and Taiwan provinces and their fluoride concentrations are high, some having fluoride as much as 15 mg/L.

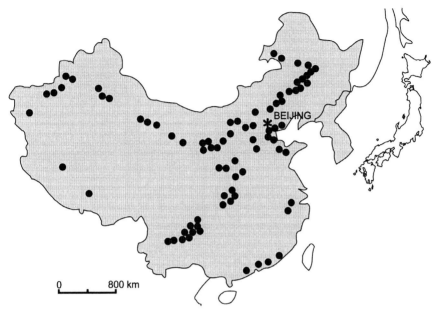

Fig. 4.19. Distribution of high groundwater fluoride contents in China (Zheng and Hong, 1988)

Fluoriferous coal is another major cause of endemic fluorosis in China. In the Guizhon Province in the south west of China, for example, more than 10 million people suffer from various forms of fluorosis (Dai et al., 2004). The rate of incidence of osteofluorosis is about 80% in some villages, and almost every family has at least one member with serious fluorosis that had resulted in disability and paralysis. The main cause of this fluorosis is the burning of clay mixed with coal. The coal in some cases has fluorine contents greater than 500 mg/kg, very much higher than the world average for coal, namely 80 mg/kg (Luo et al., 2003). Further, the villagers use clay as coal-burning additive in the furnace and as a binder in briquette-making. The ratio of the mixture of coal and clay ranges from 2.1 to 4.1. Interestingly, the clay contains fluoride concentrations ranging from 100 to 2455 mg/kg, with an average of 1027 mg/kg. Dai et al. (2004) are therefore of the view that clay is more important than coal in the incidence of fluorosis in the Guizhan Province. Figure 4.20 shows the distribution of endemic skeletal fluorosis prevalence in China.

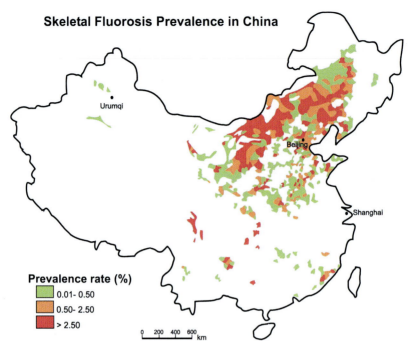

Fig. 4.20. Distribution of endemic skeletal fluorosis prevalence in China (modified from Tan et al., 1989)

Brick Tea Fluorosis in China

In addition to the water-related and the coal-related fluorosis, a third source of fluorosis is the popular drink brick tea. The brick-tea fluorosis, as it is known in China, was discovered in the 1980s. Tea has the ability to accumulate higher concentrations of fluoride and the tea drink therefore, as in some parts of China, can be a source of fluorosis.

An epidemiological survey was conducted by Cao et al. (2003) in Naqu County, Tibet, to investigate the manifestations of fluorosis in adults drinking brick tea. The fluoride concentration of the water sources was only 0.10 mg/L and there was no evidence of fluoride air pollution. The foods processed from brick tea, zamba and buttered tea however, had fluoride contents of 4.52 and 3.21 mg/L respectively. The adult daily fluoride intake was 12 mg of which 99% originated from foods containing brick tea. Of the cases studied, 74% had osteosclerosis type skeletal fluorosis. Cao et al. (2003) were of the view that brick tea-type fluorosis had even greater health impacts than the water- type and the coal-combustion type.

DEFLUORIDATION OF HIGH FLUORIDE GROUNDWATER

The presence or absence of fluoride in their water supplies is generally not known by the public, particularly in developing countries where in most areas the drinking water is obtained direct from the ground. Unlike in the case of excessive dissolved iron in the dug and deep wells, where a colour and objectionable taste is imparted, fluoride imparts neither colour nor taste. It should be noted that both Fe^{2+} and Fe^{3+} are only soluble at very low pH. Only chemical analyses can detect its presence and the concentration. This is a major reason why such a high percentage of the population living in fluoride-rich areas and who suffer from fluorosis are not even aware of the problem until the later stages.

Since the vast majority of the people affected by fluorosis in tropical countries live in rural areas without central water treatment plants, they obtain their domestic water supplies in the untreated form from lakes, rivers, surface wells and deep wells, which may contain biological and chemical contaminants detrimental to health. The need for simple water treatment techniques at the rural level therefore is a prime need for fluorosis-affected tropical developing countries.

Phantumvanit et al. (1998) who developed a defluoridator for individual households in Northern Thailand noted that the shortcomings of most defluoridation methods are:

> high cost of plants,
> high operational and maintenance costs,
> low capacity for removing fluoride,
> lack of selectivity for fluoride,
> undesirable effects on water quality,
> generation of waste that is difficult to handle and
> complicated procedures.

Based on the nature of process, defluoroidation techniques are classified under:
- (a) adsorption and ion exchange
- (b) precipitation
- (c) electrochemical methods
- (d) membrane techniques

Table 4.8 shows the various materials used in the different defluoridation techniques.

A highly efficient, simple house hold defluoridator using burnt bricks was employed in many areas with high fluorosis in Sri Lanka (Padmasiri and Dissanayake, 1995). This defluoridator is suitable for developing countries and is especially suited on account of its easy installation, maintenance, ready availability of the defluoridating raw material used, i.e., burnt bricks, and low cost, thus achieving Village Level Operation and Maintenance status (VLOM)(Figure 4.21).

Heidweiller (1990) reviewed some of the more common techniques of defluoridation of fluoride-rich water. In the Nalgonda technique, used in the fluorosis-affected Andhra Pradesh, India, a combination of alum and lime mixed with bleaching powder is added to fluoride-rich water, stirred and allowed to settle (Nawlakhe and Bulusu, 1989). The fluoride is removed by the process of flocculation, sedimentation and subsequent filtration. Among the other common defluoridation techniques are: use of activated carbon, activated alumina, ion-exchange resins, clay minerals, clay pots and bone char and gypsum (Edmunds and Smedley, 2004; Weerasooriya et al., 1989; Jinadasa et al., 1988; Weerasooriya et al., 1994; Jinadasa et al., 1991, Schuiling, 1998).

Fig. 4.21. Household defluoridator using burnt bricks distributed in villages in Sri Lanka

Table 4.8. Materials and methods for defluoridation (Mariappan and Vasudevan, 2002)

Adsorption	Ion exchange	Precipitation	Others
Carbon materials	NCL poly anion	Lime	Electrochemical
Wood	Resin	Alum	(Aluminium electrode) Electrodialysis
LigniteCoal,	Tulsion A27	Lime and Alum	Reverse Osmosis
Bone	Lewatit-MIH-59	(Nalgonda Technique)	
Petroleum residues	Amberlite IRA-400	i).Fill and Draw	
Nut shells,	Deacedodite FF-IP	ii).Continuous flow	
Paddy husk	Waso resin-14	iii).Package Treatment plant for HP	
Avaram bark	Polystyrene	Alum floc blanket-method	
Coffee husk,		Poly-aluminium chloride- (PAC)	
Tea waste		Poly-aluminium Hydroxy-sulphate (PAHS)	
Jute waste			
Coconut shell			
Coir pith,			
Fly ash			
Carbion,			
Defluoron-1			
Defluoron-2			
Activated alumina			
KRASS, Bauxite			
Serpentine			
Clay minerals			
Fish bone,			
calcite			
Bio-mass			

CHAPTER 5

IODINE GEOCHEMISTRY AND HEALTH

It has been estimated that about 29% of the world population is at risk from some form of iodine deficiency disorder. It is the world's most common cause of mental retardation and brain damage with 1.6 billion people at risk, 50 million children affected and 100,000 cretins born every year. Pharaoh (1985) considers endemic cretinism to be the most important form of IDD, even though other diseases such as still births, abortions, congenital abnormalities and impaired mental function of children are also major health problems (Stewart and Pharoah, 1996).

These iodine deficiency disorders (IDD) are particularly severe in the lands of the tropical belt. The impaired mental function of people has serious direct and indirect impacts on all aspects of life among these people (Figure 5.1). Figure 5.2 illustrates the world distribution of areas affected by IDD and it is apparent that the problem has assumed serious proportions and needs worldwide attention. Among these countries are those in the South Asian region namely Bangladesh, Vietnam, Myanmar, Indonesia, Nepal, India and Sri Lanka. Table 5.1 shows the magnitude of the problem in these countries, all of which lie in the tropical environment.

The geochemistry of iodine and its chemical species and its impact on the health of a very large population of the world is one of the most important fields of study in the field of medical geology.

THE IODINE CYCLE IN THE TROPICAL ENVIRONMENT

Figure 5.3 illustrates a schematic model for the possible transformation and geochemical pathways of iodine in the tropical environment. The sea is a major source of iodine and in lands adjacent to the sea, the marine influence will be particularly strong and the distance from the sea will therefore be of some importance. The expected decrease of iodine from the sea towards the land however is not always regular and other factors such as atmospheric circulation may play an inhibiting role.

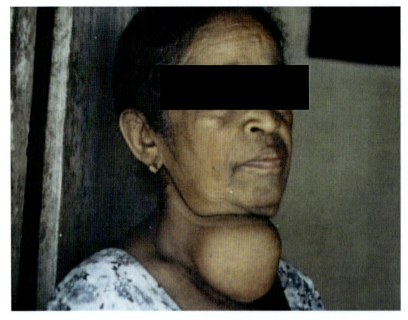

Fig. 5.1. A typical case of endemic goitre from Sri Lanka

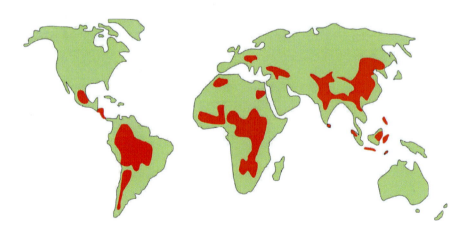

Fig. 5.2. World map showing areas affected by iodine deficiency. Other areas, especially in Africa and the Middle East, may also have iodine deficiency problems but have not been surveyed in detail (Dunn and van der Harr, 1990).

Iodine Geochemistry and Health

Table 5.1. The incidence of iodine deficiency disorders in some Asian countries (Source of data from WHO)

Country	At Risk from IDD	Goitre	Cretinism	Other Diseases
Bangladesh	37223	10230	491	2796
Bhutan	1446	993	95	704
Myanmar	14464	5700	404	2591
India	149580	54540	3338	18503
Indonesia	29772	10131	749	3569
Nepal	15056	9438	736	5145
Sri Lanka	1861	3107	140	580
Thailand	20438	7927	539	3331
Total	227740	102006	6488	36769

Population (x1000)

The high rainfall resulting in intense leaching of elements from the rocks and abundant laterite formation plays a major role in the geochemical cycle of iodine in the tropics. The presence of acid soils, organic matter and rapid groundwater flow also influence the leaching of iodine in the tropical environment to a marked degree. The most significant feature of the geochemical cycle of iodine is that the iodine abundance and mobility is most marked in the surface environment; these surface phenomena involving the soil-atmosphere interactions are of extreme importance.

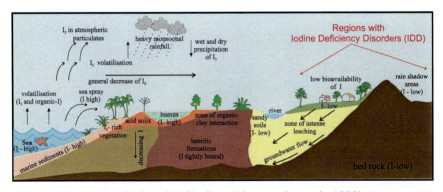

Fig. 5.3. The geochemical cycle of iodine (Dissanayake et al., 1998)

Radioactive iodine-129 (half life 15.7 Ma) has been released into the environment by nuclear weapons and the operation of nuclear facilities such as spent fuel reprocessing plants. The high mobility of iodine in the surface environment therefore becomes a critical factor in radioactive water management. The geochemical behaviour of iodine is now better understood as a result of studies of iodine radioisotopes (Muramatsu et al., 2004).

As shown in Figure 5.3, the sea which is the major source of iodine in the geochemical cycle has an average concentration of around 58 µg/L (Fuge and Johnson, 1986; Fuge, 1996). Iodate (IO_3^-) is the most stable form of iodine in sea water. It gets reduced to iodide (I^-) in surface waters by the biological action. Seaweeds and phytoplankton release iodine containing organic gases (CH_3I, CH_2I_2 etc) and these pass into the atmosphere where they undergo further chemical changes due to sunlight. The iodine in the atmosphere then migrates inland and is deposited inland by the climatic and topographic conditions (Johnson et al., 2003a). Figure 5.4 illustrates a simplified model showing part of the iodine cycle involving transport from marine to the terrestrial environment.

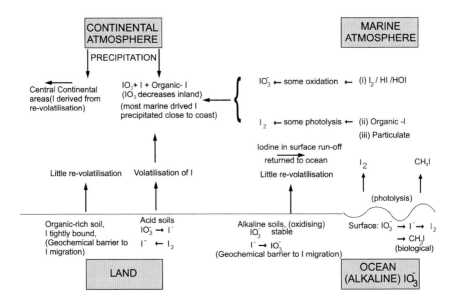

Fig. 5.4. Iodine transport from the marine to the terrestrial atmosphere (Fuge, 1996)

The distribution of iodine in the earth's crust was studied and the data compiled by Muramatsu and Wedepohl (1998). From among the magmatic and metamorphic rocks (Table 5.2), metasedimentary gneisses, mica schists and granulites have as little as 12 to 25 µg/kg I and have lost from 75 to >95% of their iodine content at metamorphic temperatures. Granites, granodiorites, tonalites, and basalts are even lower in iodine and contain 4 to 9 µg/kg I almost independent of the class of magmatic rock. Table 5.3 illustrates soil iodine levels of some parent rock materials.

Table 5.2. Iodine in magmatic and metamorphic rocks (μg/kg I ± SD) (SD, standard deviation of two to four determinations, mostly three determinations) (Muramatsu and Wedepohl, 1998)

Gneisses mica schists, amphibolites, marbles of upper continental crust	
Gneiss, Variscan Loja near Persenbeug, Lower Austria	9.8 ± 2.8
Gneiss, Variscan KTB deep hole Windischeschenbach Bavaria, Germany	38 ± 2
Gneiss, Variscan KTB deep hole Windischeschenbach Bavaria, Germany	46 ± 9
Gneiss, Variscan HTKB deep hole Windischeschenbach Bavaria, Germany	36 ± 3
Garnet sillimanite gneiss, Caledonian E Frivole near Arendal, Norway	8.2 ± 0.6
Two-mica gneiss, Alpine Brione Verzasca Valley, Switzerland	4.0 ± 0.6
Mica schist Panafrican Damara Belt, Namibia	18 ± 2
Amphibclite Panafrican Amphibolite Belt Augaigas Farm W Windhoek, Namibia	23 ± 3
Marble, Alpine Kleintal Gleinalpe Scyria, Austria	31 ± 6
Granulites of lower continental crust	
Hypersthene perthite granulite, Caledonian 1 km S Tveite Arendal, Norway	11 ± 1
Hypersthene plagioclase granulite, Caledonian 450 m N Skuggerik Arendal, Norway	10 ± 1
Hypersthene perthite plagioclase granulite, Caledonian Ferry to Tromoey Arendal, Norway	15 ± 2
Granite, granodiorites, tonalites	
Granites, Variscan Germany (composite of 14)	82 ± 3
Hydrothermally altered granites (greisens), Variscan, Germany (composite of 24)	165 ± 5
Granite, Variscan Schrems, Lower Austria	4.3 ± 1.1
Granite, Variscan Eisgarn, Lower Austria	2.0 ± 0.4
Granite, Variscan Weinsberg, Lower Austria	2.0 ± 0.4
Granite, Variscan Mauthausen, Lower Austria	2.4 ± 0.5
Granite (Rapakivi-type), Svecofennian Balmoral Rauma Bottnia, Finland	9.2 ± 1.1
Granite, Westerly RI, USA (USGS Standard G-2)	38 ± 3
Granodiorite Azuma-mura, Gumma Japan (GSJ Standard JG-1)	5.0 ± 1.8
Granodiorite trondhjemite, Variscan Melibokus Odenwald, Germany	9.9 ± 1.0

Granodiorite Silver Plume Ouarry, CO USA (USGS Standard GSP-1)	15 ± 2
Tonalite, Alpine Melirolo S Chiavenna, North Italy	7.8 ± 0.8
Tonalite, Variscan Rastenberg, Lower Austria	4.4 ± 1.1
Dacitic rhyolite, Medicine Lake, CA/OR, USA	35 ± 6
Rhyolite (obsidian), Wada Toge, Nagano Prefecture, Japan (GSJ Standard JR-2)	69 ± 11
Andesites	
Andesite, Manazuru-machi, Hakone, Japan (GSJ Std. JA-1)	17 ± 4
Andesite, Tumago-mura, Gunma, Japan (GSJ Standard JA-2)	4.2 ± 0.4
Andesite, Guano Valley, OR, USA (USGS Standard AGV-1)	59 ± 9
Andesite, Medicine Lake, CA/OR, USA	13 ± 3
Basalts	
Quartz tholeiite, Tertiary Borken Hessian Depression, Germany	7.4 ± 0.5
Alkali olivine basalt, Tertiary Bramburg Goettingen	5.8 ± 0.5
Alkali olivine basalt, (basanitite), Tertiary Rhuender Berg Hessian Depression, Germany	8.8 ± 2.9
Olivine nephelinite (melilite bearing) Tertiary Westberg Hofgeismar Hessian Depression, Germany	14 ± 5
Olivine nephelinite (melilite bearing) Tertiary Hoewenegg Hegau, Germany	10 ± 1
Peridotites	
Spinell Iherzolite, Balmuccia Ivrea Belt, North Italy	11 ± 2
Spinell Iherzolite, Baldissero Ivrea Belt, North Italy	12 ± 2

Muramatsu and Wedepohl (1998) have given values of 119 µg/kg, 777 µg/kg and ~300 µg/kg I respectively for the continental crust, oceanic crust (including seawater) and the bulk Earth's crust. Nearly 70% of I is thought to be found in ocean sediments. From among the sedimentary rocks and organic matter (Table 5.4), the high average concentrations of 30 mg/kg I in deep-sea carbonates and 2.5 mg/kg in continental limestones had been accumulated by planktonic and shallow sea organisms, respectively. Deep-sea shales and continental shales carry high concentrations of iodine.

Table 5.3. Summary of statistics for soil iodine contents classified by parent material (in mg/kg) (Johnson, 2003a)

Parent Material	Number	Min.	Max.	Mean	Geo. mean
Alluvium	157	0.1	56.5	3.56	1.28
Carbonates	117	0.1	22.6	4.38	3.05
Other Sedimentary	157	0.06	38.7	4.58	2.00
Igneous Extrusive	114	0.1	72	14.16	6.31
Igneous Intrusive	21	0.4	83.2	10.66	3.75
Metamorphic	41	0.1	21	3.37	1.15
Peat	4	11.6	68.4	32.9	26.52
Sand	32	0.1	9.8	1.56	0.71

Table 5.4. Iodine in sedimentary rocks and organic matters (µg/kg ± SD) (Schnetger and Muramatsu, 1996, Muramatsu and Ohmoto, 1986, and Muramatsu et al., 1983)

Pelagic Sediments	
Clay, Atlantic Ocean (16°54' N, 59°31' W) 5 m sediment depth, 5030 m water depth (A 160, 8) ≤ 1% CaCO$_3$	4700 ± 108
Clay, Atlantic Ocean (19°05' N, 59°42' W) less than 6 m sediment depth, 5300 m water depth (A 160, 7) ≤ 1% CaCO$_3$	1660 ± 167
Clay, Pacific Ocean (20°49' N, 125°0.5' W) 4530 m water depth (Y2680 2P 831, 25) 3% CaCO$_3$	2680 ± 215
Clay, Pacific Ocean (14°55' N, 124°12' W) 8.2 m sediment depth, 4270 m water depth (50BP 240) 2.6% CaCO$_3$	5330 ± 135
Clay, Atlantic Ocean (16°54' N, 59°31' W) 2.65 m sediment depth, 5030 m water depth (A 160, 8) 29% CaCO$_3$, 0.22% C	5240 ± 410
Foraminifera ooze, Pacific Ocean (15°39' S, 114°18' W) 3221 m water depth (42 HG 84) 81% CaCO$_3$, 0.28% C	20000 ± 340
Shales, Corg –rich shales	
Shales, marine Paleozoic, W Europe (composite of 36) 2% CaCO$_3$, 1.9% C, 0.24% S	827 ± 22
Shales, marine Paleozoic, Japan (composite of 14) 1.3% CaCO$_3$, 0.4% C, 0.12% S	4520 ± 85
Shale, marine Triassic, Friedland near Goettingen Germany	758 ± 81
Shale, Corg-rich Archean Fig Tree, South Africa (Capetown Fg 14)	248 ± 20
Shale, Corg-rich Upper Cambrian Oslo Fjord, Norway (Pr 631)	510 ± 21
Shale, Corg-rich Lower Ordovician Kopingbland, Sweden (Pr 635)	2110 ± 207
Shale, Corg-rich Upper Permian Kupferschiefer West Drente, Netherlands (Pr 785) 4% C	413 ± 23
Shale, Corg-rich Upper Permian Kupferschiefer Calberlah Braunschweig, Germany (Pr 724) 7.4% C	6150 ± 230
Shale, Corg-rich Upper Permian Kupferschiefer Eisleben, Germany 6.9% C, 16% CaMg(CO$_3$)$_2$	4840 ± 42
Shale, Corg-rich, Etzel near Bremen, Germany, 16% C	526 ± 47

Shale, Corg-rich Lias-ε Hohenassel near Hildesheim, Germany (R3/300 m) 6.5% C	2970 ± 293
Shale, Corg-rich Lias-α Levin brickyard Goettingen, Germany	195 ± 3
Shale, oil-bearing Green River Formation, USA (USGS Standard SGR-1) 3.2% C, 20% $CaCO_3$	334 ± 45
Greywackes, sandstones	
Greywackes, Paleozoic Central Europe (composite of 17) (Pr 1025) 4% $CaCO_3$, 0.1% S	168 ± 20
Greywacke Lower Carboniferous Langelsheim Harz Mountains, Germany (B 18)	103 ± 7
Greywacke Lower Carboniferous Andreasbach Harz Mountains, Germany (5IIIb)	63 ± 1
Greywacke Lower Carboniferous Clausthal Silberhuette Harz Mountains, Germany (C 4)	138 ± 2
Greywacke Lower Carboniferous Soesetal Harz Mountains, Germany	52 ± 5
Greywacke Lower Carboniferous Gr. Steimkertal Harz Mountains, Germany (Pr 683)	80 ± 6
Quartz sandstones, Carboniferous Germany (composite of 11) (Pr 680) 80% quartz, 2% $CaCO_3$	123 ± 11
Quartz sandstones, Lower Triassic Germany (composite of 23) (Pr 910) 65% quartz, 2% $CaCO_3$	144 ± 11
Quartz sandstones, Cretaceous Germany (composite of 11) (Pr 1014)	
Limestones	
Limestones, Devonian Germany (composite of 32) 0.1% S	1610 ± 127
Limestones, Middle Triassic Reckershausen Goettingen, Germany (Pr 900)	260 ± 47
Limestones, Middle Triassic Dransfeld near Goettingen, Germany (Pr 901)	792 ± 154
Limestones, Jurassic Germany (composite of 45) 0.17% S	3370 ± 275
Limestones, Upper Jurassic W Eroden, Germany (Pr 983)	1970 ± 83
Limestones, Upper Jurassic Solnhofen, Bavaria, Germany (Pr 984)	3870 ± 79
Limestones, Cretaceous Germany (composite of 16) 0.1% S	1940 ± 96
Organic matter, cool	
Orchard leaves, dry (Standard NIST 1571)	215 ± 64
Brown algae, *Hijikia fushiforme*, Japan (mean of four dry samples)	490,000 ± 126,000
Brown algae, *Undaria pinnatifida*, Japan (mean of two dry samples)	87,000 ± 64,000
Red algae, *Gloiopeltis furcata*, Japan (mean of five dry samples)	100,000 ± 14,000
Red algae, *Porphyra tenea*, Japan (one dry sample)	44,000
Oyster tissue, dry (Standard NIST 1566a)	4100 ± 300
Hard coal, Queen Luise Mine Upper Silesia, Poland	4210 ± 1615

From a medical geology point of view, the iodine status of soils is most important. The amount of soil and its ability to retain it are two factors that need consideration in the study of the geochemical pathways of iodine. As illustrated in Figure 5.5, Fuge and Johnson (1986) discussed three main forms of iodine in the soil, namely (a) mobile iodine (b) insoluble iodides (c) fixed iodine. The property of the soil which fixes the iodine was termed Iodine Fixation Potential (IFP).

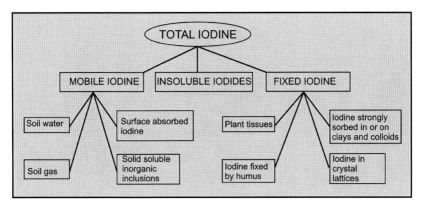

Fig. 5.5. Suggested forms of iodine in soils (Fuge and Johnson, 1986)

The IFP is particularly important for tropical soils since Fe-Mn and Al oxides are abundant in such soils and these have the ability to fix iodine strongly. The organic matter in the soil also absorbs iodine strongly and the bioavailability of iodine may therefore be relatively small.

In view of the fact that the tropical environment is characterized by laterite formation and in which Fe, Mn, and Al oxides are very common, the iodine geochemistry is markedly influenced by these minerals. Whitehead (1974) has shown that the sorption of iodine by aluminium and iron oxides is markedly influenced by pH, greater sorption in acidic conditions and no sorption under neutral conditions. Johnson et al. (2003a) and Johnson (2003b) compiled global soil data for iodine and were of the view that the geometric mean for the iodine levels in soil is 3.0 µg/g. On a textural classification for soils, the following order for mean values of I was determined. Figure 5.6 illustrates the distribution of the reported iodine results.

Peat (7.0)> clay (4.3)> silt (3.0)> sand

In studies pertaining to the effect of iodine concentration on IDD, it is of extreme importance to note that it is the bioavailable iodine and not the total iodine in the soil that would influence the incidence of IDD. Generally less than 10% of the soil iodine is known to be extracted with cold water and this is considered as a good indication of the bioavailability of iodine.

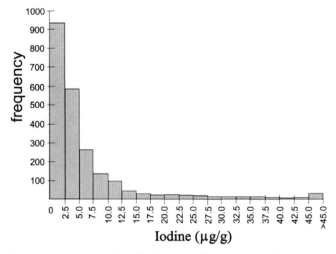

Fig. 5.6. Histogram showing the distribution of reported iodine concentrations in soil (Johnson, 2003)

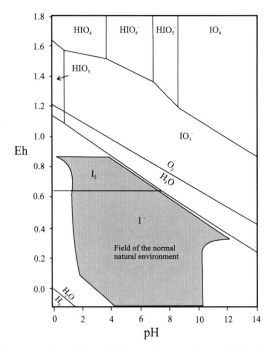

Fig. 5.7. Eh-ph diagram for iodine (Vinogradov and Lapp, 1971)

From among the chemical species of iodine, iodide is the most mobile form in the soil and which is easily available to plants. Iodate is relatively less mobile. Acidic soil conditions are known to favour iodide and the alkaline oxidizing conditions prefer the less soluble iodate form. Figure 5.7 illustrates the Eh-pH diagram for iodine.

Iodine sorption on clays and humic substances

In the tropical soils, clay minerals and humic substances are two of the important iodine fixers. Since much of the population of the developing countries in the tropical belt live in close association with the immediate physical environment, the IFP of soils needs to be carefully considered in epidemiological studies. Organic matter, mostly humic substances however display a much higher IFP than clay.

Hamid and Warkentin (1967) studied ^{131}I as a tracer for water movement in soils and observed that iodine is adsorbed onto clay particles. This observation was also made by Vinogradov and Lapp (1971). In an experiment conducted, De et al. (1971) added iodide solutions of varying temperature to soil clays and observed that only the clay minerals took up iodide with illite adsorbing more iodide than kaolinite or montmorillonite.

Humic substances in the environment are known to play a major role in the speciation and geochemical mobility of chemical elements. The conversion of chemical species into toxic or non-toxic forms has important implications on the health of individuals living in a particular geochemical habitat. Dissanayake (1991b) described the association of certain metals in the environment with organic groups of humic substances that had been studied from the point of view of the incidence of some geographically distributed diseases such as cancer, where selenium and molybdenum appear to play an important role.

It was shown in Figure 5.5 that iodine is strongly fixed by humus, and soils rich in humus therefore tend be enriched in iodine but with low bioavailability depending on pH conditions. As noted by Johnson et al. (2003a), contrary to what might be expected, organic-rich soils, though high in iodine content, do not provide much iodine to the food chain in view of its strong fixation to humus and hence low bioavailability.

The nature and the mechanism of the fixation of iodine within the humus structure is not well known, but the sizes of the iodine ion and its oxyanion are perhaps of special importance in this process.

For the humic compounds which comprise the bulk of naturally occurring organic matter in soil and water, a discrete structure cannot be given. Among the many different functional groups present are carboxyl, phenolic, enolic hydroxyl, alcoholic hydroxyl, quinone, hydroxyquinone, carboxyl, ester lactone, ether and amino groups (Rashid and King, 1970). It is generally accepted that the major oxygen containing functional groups are carboxyl (-COOH), hydroxyl (-OH) and carbonyl (>C = 0).

At low pHs (e.g. pH <4) absorption of a negative ion such as I^- or IO_3^- could perhaps be brought about by a reaction such as that shown in the lower part of Figure 5.7 (humic structure) where the I^- or IO_3^- ions become bound to the positive ion.

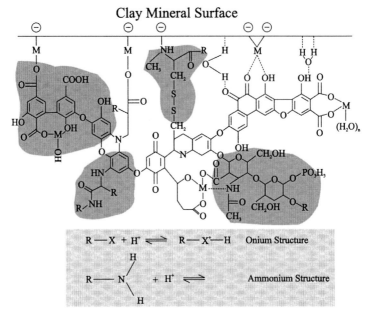

Fig. 5.8. Postulated structure of humic material in association with clay A- lignin, B- peptide C- cellulose or chitin M- Metal (Koss, 1977) Iodine species could perhaps be bound to the onium and ammonium structures (Dissanayake et al., 1998)

Figure 5.8 illustrates a postulated structure for humic materials and its bonding with the surface of clay. Such humus-clay associations as found in tropical soils may function as good loci for the adsorption of iodine. The organic matter-clay interactions in tropical soils and their implications on the geochemical cycle of iodine is therefore worthy of detailed investigation.

Effect of Microbial Activity on Iodine Geochemistry

Several studies have shown that microbial activity may play an important role in the iodine cycle. This is of special importance to tropical countries, notably with flooded soil such as in the rice fields, where microbial activity may be intense. Razaq et al. (1987) had shown that in calcareous soils, some part of the anionic iodine is converted to molecular I_2 through microbial or enzymatic processes or reactions with by-products. It was reported by Higgo et al. (1991) that microorganisms play a role in the sorption of both I^- and IO_3^-. Behrens (1982) had shown that microorganisms are involved in the loss of I^- from fresh water aquatic systems and it was suggested that the reactions were extra cellular, possibly enzymatic oxidation of I^- to I_2, which then reacts with organics probably proteins. Sheppard and Hawkins (1995), however, were of the view that microbes may play only a minor and indirect role in iodine sorption through the decomposition of organic matter.

Muramatsu and Yoshida (1999) studied the effects of microorganisms on the fate of iodine in the soil environment. They observed that the behaviour of iodine in the soil environment was influenced by microbial activities, perhaps their products (e.g. enzymes) as well, both in the sorption and desorption processes. In the case of flooded soils (e.g. rice paddy soils) iodine was desorbed due to the reducing conditions (low Eh) created by the effects of soil microorganisms. It is expected that evaporation of biogenerated methyl iodide from the soil-plant system may also result in lowering of iodine levels in soils, specifically in rice fields and marshes. Muramatsu and Yoshida (1999) observed that the influence of soil microorganisms is significant for both stable and long–lived radioactive iodine in the environment.

They incubated a number of soil samples collected from paddy fields, farms and forests, with radioactive iodine tracer. Where the samples were incubated with the antibiotics streptomycin and tetracycline, which are specific inhibitors for prokaryotes (mainly bacteria), iodine volatilization ceased. When cycloheximide, a specific inhibitor for eukaryotes (filamentous fungi and yeast), was added, there was no significant change, indicating that soil bacteria contribute to iodine volatilization from soil environments.

From among the isolated 100 bacterial strains from a variety of environments, it was found that about 40% of these strains showed significant CH$_3$I production, with some strains showing very high (1-5% of the total) production of CH$_3$I.

Iodine in Drinking Water

Water contains the more mobile form of iodine, thus, the iodine content of water is a good index of the iodine status of the environment (Johnson et al., 2003b). However drinking water does not represent a major source of iodine (Fig 5.9) (iodine in air, water etc), even though it is more amenable to chemical analysis. A threshold value of 3 µg/L has been given as a marker value for iodine deficient environments. Depending on a variety of environmental factors, the iodine contents in drinking water markedly varies. Results reported so far reveal a range from <0.1 to 150 µg/L with the average being 4.4 µg/L (Johnson et al., 2003b). It is known that deep-water sources are richer in iodine than surface waters. Drinking water provides only about 10% of the Recommended Dietary Allowance (RDA), i.e. about 150 µg per day for iodine. In areas, particularly in rural parts in developing countries of the tropical belt, where food is obtained only from the immediate physical environment and where iodine supplementation is not available, water my contribute a greater iodine input of ~20%.

Iodine in Food

Since iodine is not an essential element for plants there is no direct correlation between iodine in soil and its content in the crops. Excessive iodine however is toxic to plants as shown by the occurrence in rice of 'Reclamation Akagare' disease, a physiological disorder caused by flooding paddy fields on iodine-rich soils (Johnson et al., 2003b). Table 5.5 shows the iodine contents of some foods.

Iodine Geochemistry and Health

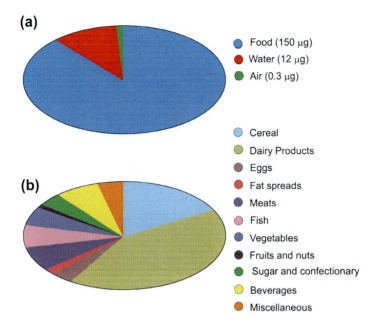

Fig. 5.9. (a) Diagram showing the estimated relative proportions of food, water and air to the daily iodine intake of a person living in a developed country. (b) Chart showing the relative contribution of iodine to the diet from different foods (Data from U.K. National Diet and Nutrition Survey) (from Johnson et al., 2003b; reproduced with kind permission from the British Geological Survey)

Table 5.5. Iodine content in some foods (http://www.healthyeatingclub.com)

Food	Iodide content (µg/100 g food)
Iodized salt	3000
Sea food	66
Vegetables	32
Meat	26
Eggs	26
Dairy products	13
Bread and cereals	10
Fruits	4

The iodide content of foods and total diets vary depending on geochemical, soil and other conditions. The major natural food sources for iodine are marine fish, shell fish, marine algae and seaweeds, and dairy products. Arable crops contain less than 50 µg/kg (fresh wt.) iodine in the order, legumes >vegetables >fruit. Since iodine is not mobile in plants, it is not concentrated in seeds. Grain crops such as rice are poor in iodine and since rice is the staple diet in many tropical countries, this observation is of special importance.

In those countries where sea food is consumed more, the general iodine intake is high bearing in mind that iodine can average 1000-2000 µg/kg (fresh weight) in some fish.

Plate Tectonics, High Altitudes and Iodine Cycling

It has been observed that in many mountainous regions of earth there appears to be a deficiency of iodine and an increased incidence of iodine disorders (Stanbury and Hetzel, 1980). High altitudes are considered as special domains with characteristic climate, soil, trace element deficiencies, human and animal health. Evidence has accumulated that populations at high altitude are prone to develop essential element deficiencies as exemplified by iodine and selenium (Iyengar and Ayengar, 1988).

In medical geology, the factors related to the occurrence of deficiencies and excesses are broadly classified as:

(a) Geochemical components in the environment, most notably in the soil. These may have a stronger influence in the soil-plant-animal-human health food chain.

(b) The bioavailability of trace elements in a given diet. This can be taken to mean the fraction of the trace elements in the diet that is absorbed by the gastrointestinal tract and which is available for metabolism.

(c) Inter-element interactions. Some elements such as calcium, lead and zinc, when in excess in the diet may interfere with the absorption of zinc, iron and copper (Underwood, 1977).

(d) The health status of the subjects per se. Special circumstances. e.g. life at high altitudes, entailing living under complex stress and related environmental constraints (Iyengar and Ayengar, 1988).

Johnson et al. (2003b) studied the levels of iodine and IDD in the mountains of Morocco. They chose the Ounein Valley in the Atlas Mountains as a case study following an earlier study by sociologists and nutritionists.

The remote mountainous area lay ~150 km inland from the Atlantic coast. Agadir, an area on the coast was selected as a control study. As shown in the box and whisker plots (Figure 5.10), Ounien valley had very low iodine in both surface waters and soils. As expected, IDD was higher in the Ounein valley located in the high altitude area than in the Agadir area.

It is worthy of note that altitude is not necessarily a causal factor in IDD. Non-mountainous areas are also known to have a high incidence of IDD. However it is the effect of the high altitude on the geochemical pathways of iodine that should be taken into account. The geochemical cycle of iodine as illustrated in Figure 5.3 shows why some mountainous areas are seriously deficient in iodine.

The theory of plate tectonics figures prominently in the iodine cycle of the crust and mantle. The iodine pathways within this cycle have an impact on the incidence of IDD, when the geological and geographical locations of the habitats concerned are taken into account. Stewart (1990) commented on the influence of plate tectonics in iodine cycling and noted IDD belts occurring in Asia. Kelly and Snedden (1958) and Kochipillai et al. (1980) also suggested such occurrences in Papua New Guinea and Myanmar.

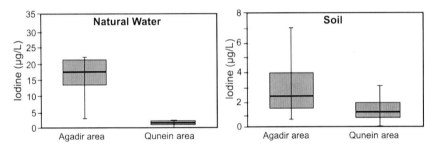

Fig. 5.10. Total iodine content in water and soils of Agadir and Ounein regions of Morocco (source: Johnson et al., 2003b; reproduced with kind permission from the British Geological Survey)

In their very recent study, Muramatsu et al. (2004) investigated the recycling of iodine in the subduction zone at the Chiba prefecture of Japan which produces one third of the world's iodine from brines. Here the iodine concentrations are in excess of 100 mg/L and typically contain methane (Figure 5.11). The Kazusa Group, which was the host formation of the brines, consisted of marine sediments of Pliocene to Pleistocene age. Here the average I concentration was more than 2000 times higher than that of seawater. Comparison with sea water showed that those elements which are accumulated in marine sediments (e.g. I, Mn, Ba) were the most enriched in the brines.

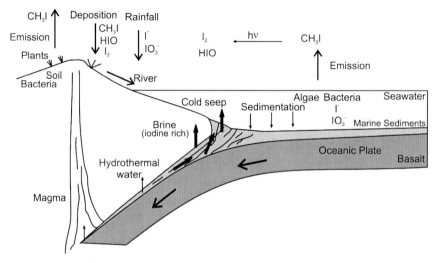

Fig. 5.11. Recycling of iodine in subduction zone (after Muramatsu et al., 2004)

Muramatsu et al. (2004) showed that iodine enrichment in the brines was caused by remobilization from subducted marine sediments associated with the release of pore waters in the fore-arc area. Similar enrichments of iodine had also been observed in other fore-arc areas such as those of New Zealand and Trinidad.

Since the largest reservoir of crustal iodine is found in marine sediments where it is closely associated with organic material, and since the cosmogenic radioisotope ^{129}I is long lived, this isotopic system could be used to study sediment recycling in subduction zones. Snyder and Fehn (2002) studied the ^{129}I/I ratios in volcanic fluids from four geothermal centres and a number of crater lakes, fumaroles, hot springs and surface waters in Costa Rica, Nicaragua and El Salvador.

The iodine ages indicated that the magmatic end member for the volcanic fluids originates in the deeper parts of the subducted sediment column, with small additions from older iodine mobilized from the overlying crust. The higher concentrations of iodine in geothermal fluids, combined with iodine ages demonstrated that remobilization in the main volcanic zone (and probably also in the fore-arc area) is an important part in the overall marine cycle of iodine and similar elements (Snyder and Fehn, 2002).

IODINE AND HEALTH

Iodine is needed by the human body for the synthesis of thyroid hormones by the thyroid gland. The bioavailable iodide ion is obtained from food and water and dietary iodine is converted into the iodide ion before absorption by the gastrointestinal tract, and all biological actions of iodide are attributed to thyroid hormones. The thyroid cells are the only cells in the body which can absorb iodine and the metabolic role played by the thyroid gland therefore is extremely important, considering the fact that every cell in the body depends on thyroid hormones for regulation of their metabolism.

The thyroid gland located in the front part of the neck is the biggest gland in the neck. It has two lobes-one left and the other right and wraps around the trachea joined by a narrow band termed the isthmus. The only function of the thyroid gland is to produce the thyroid hormone which is known to regulate the metabolism of the body.

The thyroid hormones made by the thyroid gland are thyroxine (T_4) and Triiodothyronine (T_3) by the process of the thyroid cells combining iodine and the amino acid tyrosine. These hormones then enter the blood stream and are circulated throughout the body. It has been estimated that the normal thyroid gland produces about 80% T_4 and 20% T_3. Triiodothyronine hormone (T_3) however is known to have about 4 times the strength of T_4.

An important gland termed the pituitary gland located at the base of the brain has a control on the thyroid gland. When T_3 and T_4 levels drop, the pituitary gland produces the Thyroid Stimulating Hormone (TSH) which then stimulates the thyroid gland to produce more hormones. After this takes place, the pituitary gland decreases its TSH production, thereby maintaining an equilibrium. The pituitary gland however, does not act alone. It is regulated by another gland termed the hypothalamus, which is a part of the brain. It produces the TSH Releasing Hormone (TRH) which stimulates the pituitary gland to release TSH.

IODINE DEFICIENCY DISORDERS (IDD)

Iodine deficiency is known to occur when the iodine intake is lower than the recommended levels. WHO, UNICEF and ICCIDD (WHO, 2001) have recommended that the daily intake of iodine should be:

90 µg for pre-school children (0 to 59 months)
  120 µg for school children (6 to 12 years)
  150 µg for adults (above 12 years)
  200 µg for pregnant and lactating women

When the mean daily intake of iodine is less than 25 µg the thyroid may no longer be able to synthesize sufficient amounts of thyroids hormones. The resulting low level of thyroid hormones in the blood (hypothyroidism) is the main factor responsible for causing damage to developing brain and other harmful effects known collectively as Iodine Deficiency Disorders (Table 5.6). WHO (2001) in the document on "The assessment of iodine deficiency disorders and monitoring their elimination" states:

Table 5.6. The Spectrum of the Iodine Deficiency Disorders (IDD) (WHO, 2001)

Fetus	Abortions
	Stillbirths
	Congenital anomalies
	Increased prenatal mortality
	Increased infant mortality
	Neurological cretinism:
	Mental deficiency, deaf mutism, spastic diplegia squint
	Myxoedematous cretinism:
	Dwarfism, hypothyroidism
	Psychomotor defects
Neonate	Neonatal hypothyroidism
Child & adolescent	Retarded mental and physical development
Adult	Goitre and its complications
	Iodine-induced hyperthyroidism (IIH)
All ages	Goitre
	Hypothyroidism
	Impaired mental function

"Iodine deficiency through its effects on the developing brain has condemned millions of people to a life of few prospects and continued underdevelopment. On a worldwide basis, iodine deficiency is the single most important preventable cause of brain damage. People living in areas affected by severe IDD may have an intelligence quotient (IQ) of up to about 13.5 points below that of those from comparable communities in areas where there is no iodine deficiency (Bleichrodt and Born, 1994). This mental deficiency has an immediate effect on child learning capacity, women's health, the quality of life of communities and economic productivity".

Since most iodine absorbed in the body finally appears in the urine, urinary iodine excretion is a good marker of very recent dietary iodine intake. The assessment of a population's iodine nutrition can be made by a profile of iodine concentrations in urine, and is the most practical biochemical marker for iodine nutrition (Table 5.7).

Table 5.7. Epidemiological criteria for assessing iodine nutrition based on median urinary iodine concentrations in school-aged children (source WHO, 2001)

Median Urinary Iodine (µg/L)	Iodine Intake	Iodine Nutrition
<20	Insufficient	Severe iodine deficiency
20-49	Insufficient	Moderate iodine deficiency
50-99	Insufficient	Mild iodine deficiency
100-199	Adequate	Optimal
200-299	More than adequate	Risk of iodine-induced hyperthyroidism within 5-10 years following introduction of iodized salt in susceptible groups
>300	Excessive	Risk of adverse health consequences (iodine-induced hyperthyroidism, autoimmune thyroid diseases)

When the thyroid gland is enlarged it is termed 'goitre'. The term 'non-toxic goitre' refers to the enlargement of the thyroid which is not associated with overproduction of thyroid hormone or malignancy.

The cause of goitre can be due to (a) deficiency in iodine intake and bioavailability as seen in many developing countries of the tropical belt with marked malnutrition and (b) an increase in TSH as a result of a defect in normal hormone synthesis within the thyroid gland. Goitre is graded according to the WHO classification shown below:

Grade 0 –   No palpable or visible goitre
Grade 1 –   A goitre that is palpable but not visible when the neck is in the normal position (i.e. the thyroid is not visibly enlarged). Thyroid nodules in a thyroid which is otherwise not enlarged fall into this category.

Grade 2 –   A swelling in the neck that is clearly visible when the neck is in a normal position and is consistent with an enlarged thyroid when the neck is palpated.

Hypothyroidism is a condition in which the body lacks sufficient thyroid hormones. It is caused by:

(i)   Result of a previous or present inflammation of the thyroid gland. A high percentage of the cells of the thyroid are damaged or dead and this affects the production of sufficient hormone. A common cause of thyroid gland failure is autoimmune thyroditis, also termed Hashimoto's thyroditis. This is related to the immune system of the patient.

(ii)   Medical treatments such as surgical removal of a part or all of the thyroid gland.

Hyperthyroidism, on the other hand, is the condition caused by the overproduction of thyroid hormones.

Endemic cretinism

Endemic cretinism is considered to be the most important iodine deficiency disorder (Pharoah, 1985). There are two types of endemic cretinism that are considered even though they may be the end members of a clinical spectrum. These are (i) neurological forms and (ii) myxoedematous forms (McCarrison, 1908). Among the characteristic features of neurological endemic cretinism, in its fully developed form, are severe mental deficiency accompanied by squint and deaf mutism, motor spasticity with disorders of the arms and legs of a characteristic nature. The myxoedematous type shows less severe mental retardation than the neurological type, but shows severe hypothyroidism, extreme growth retardation and incomplete maturation. The latter type is particularly common in Zaire whereas the neurological type is more commonly seen in mountainous areas with IDD. Since thyroid hormones are associated with cognitive and motor measures and are related to brain development, IDD are of special importance in human health (Stewart and Pharoah, 1996).

Goitrogens

Goitrogens are substances capable of producing thyroid enlargement by interfering with thyroid hormone synthesis. Goitrogens can be biological or mineralogical substances and research is being carried out on the possible goitrogens and their effect on the thyroid.

There are many naturally occurring agents that could function as goitrogens in man. Animal and in-vitro tests have shown that they possess antithyroid effects. Among the chemical groups which contain these compounds are sulphurated organics (such as thiocyanate, isothyanate, goitrin and disulphides), flavonoids, polyhydroxyphenols and phenol derivatives, pyridines, phthalate esters and metabolites, PCB's and PBB's, other organochlorines (e.g. DDT) and polyaromatic hydrocarbons. Goitrogens were classified by Gaitan (1990) into:

        (a)   agents acting directly on the thyroid gland
        (b)   agents causing goitre by indirect action.

The former group was further subdivided into (a) those inhibiting the transport of iodide into the thyroid (e.g. thiocyanate and isothyocyanate) (b) those acting on the intrathyroidal oxidation and organic binding process of iodide (c) those interfering with proteolysis, dehalogenation and hormone release.

The indirect goitrogens were thought of as increasing the rate of thyroid hormone metabolism. Goitrogens are effective particularly when the iodine intake is low and when goitrogens are continually ingested over a long period. Some foods such as cassava (manioc), cabbage, Brassica *sp.* among others are known to possess goitrogenic properties. Cassava in particular has been subjected to many investigations for its goitrogenic activity (Bourdoux et al., 1978; Delange et al., 1976).

Djazuli and Bradbury (1999) studied the cyanogen content of cassava roots and flour in Indonesia. They noted that 30 samples of cassava starch and other specialized products had a mean cyanogen content of only 5 mg/kg, whereas 29 samples of cassava flour, chip and gaplek had a much higher mean cyanogen content of 54 mg/kg. The WHO safe value for cassava flour is 10 mg/kg whereas in Indonesia it was 40 mg/kg. The study of Cassava is of special importance for the tropics since it is the third most important food source after rice and maize.

Badly prepared cassava is known to yield very high contents of cyanogenic glucosides and is converted to thiocyanate after ingestion and which in turn competes with iodide for entry into the thyroid. Its effect is particularly marked where there is an existing iodine deficiency. Peterson et al. (1995) showed that in the Central African Republic, where cassava is the main staple crop, improved cassava processing could reduce the IDD.

Geochemical goitrogens are those materials found in the geological environment, particularly in the tropics and which make the bioavailability of iodine significantly low. The availability of environmental iodine to the diet depends on several factors such as soil chemistry and soil physics, input from atmosphere, humic substances, Al and Fe oxides and clay. These materials, depending on the existing physico-chemical conditions may become geochemical goitrogens. As shown by Stewart and Pharoah (1996), in the presence of a goitrogen, iodine will act only as a limiting factor and it is the iodine to goitrogen ratio that acts as the determinant of the outcome.

Stewart et al. (2003) cautioned against the direct correlations of the incidence of IDD with environmental iodine. At present, the argument for environmental iodine deficiency is:

- Biochemical iodine deficiency is the immediate cause of the disorders.
- The source of dietary iodine is the diet
- The diet depends on the environment.
- Where IDD occurs, particularly in a community that lives close to the land, the local environment must be deficient in iodine.

Stewart et al. (2003) considered this argument to be too simple. Their study of endemic goitre in England and Wales showed that there is a lack of the expected correlation between the distribution of environmental iodine and the presence or absence of endemic goitre. This indicates that many factors, notably the activity of geochemical goitrogens complicate the direct application of correlation between IDD and environmental iodine.

As mentioned earlier in this chapter, the organic matter and clay content of soils are highly significant in such correlative studies. Shinonaga et al. (2001) carried out an important study that highlights the complex factors involved in such correlations.

These authors studied the concentrations of iodine in cereal grains cultivated at 38 locations in Austria from cereal producing sites in an agriculture area. They determined the soil to grain transfer factors (TF), which are known to vary depending on the plant species, soil characteristics, chemical forms of iodine and climatic conditions.

As shown in Table 5.8, the TFs correlated positively with iodine concentrations in cereal grains ($r = 0.70$, $p<0.001$) and a relatively good negative correlation ($r = -0.68$, $p<0.001$) was found between TF and iodine in soil. The correlation coefficient between iodine contents and clay contents in soil obtained was $r = 0.75$ ($p<0.001$). It was found that the larger the amount of clay minerals in the soil, the higher was the iodine concentrations in the soil. There was however no clear relationship between TF and organic matter and also TF and soil pH.

As mentioned earlier, humic substances are important in the iodine fixation in soil. Cooksey et al. (1985) showed that the associated constituents of humic substances (HS), resorcinol, orcinol, phloroglucinol, pyrogallol, 3, 4- and 3, 5-dihydroxybenzoic acids were potent inhibitors of the thyroid peroxidase enzyme and of thyroidal ^{125}I uptake and/or its incorporation into thyroid hormones using thyroid slices. Resorcinol was found to be goitrogenic in rats and interestingly it was found in the water supply of an endemic goitre district in western Colombia.

Recently it has been show that selenium deficiency may have significant effects on thyroid hormone metabolism and possibly on the thyroid gland itself. Here the function of type 1 deiodinase (a selenoprotein) is impaired. This selinoenzyme converts the prohormone T_4 to the active hormone T_3 which is important in some metabolic activities of the foetus, neonate and child. Selenium deficiency also leads to a reduction of the Se-bearing enzyme glutathione peroxidase which detoxifies H_2O_2 present in the thyroid gland. Reduced detoxification is known to lead to thyroid cell death.

Selenium geochemistry is therefore closely linked to the medical geochemistry of iodine and many investigations are now being carried out on the influence of selenium on IDD. It is worthy of note however, that while selenium deficiency many have a bearing on IDD, the reverse is not necessarily true. Experiments on the sorption of iodine on other minerals such as illite, goethite, clinochlore, calcite, limonite, biotite, pyrite, magnetite and hematite were carried out by Fuhrmann et al. (1998). Figure 5.12 shows the percentage of ^{125}I removed from solution over a period of

15 days in contact with powdered minerals. Pyrite and illite showed major uptake of the tracer. The pyrite removed almost all the ^{125}I from solution over 15 days, while 82% sorbed in less than 1 day. The illite (in shale) sorbed at a slower rate, with about 70% sorption after 15 days.

Table 5.8. Correlation matrix (Pearson correlation, n = 38): relation between (a) transfer factor and parameters of cereal grain and soil and (b) iodine in soil and parameters of soil (Shinonaga et al., 2001)

Parameter	Transfer Factor of Iodine	P
(a)		
Iodine content in cereal grain	0.703	<0.001
Iodine content in soil	-0.677	<0.001
Clay content in soil	-0.781	<0.001
Organic-carbon content in soil	-0.415	<0.01
pH value of soil	-0.372	<0.03
(b) Iodine in soil		
Clay content in soil	0.750	<0.001
Organic-carbon content in soil	0.520	<0.001
pH of soil	0.377	<0.02

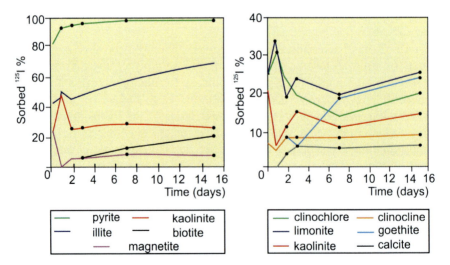

Fig. 5.12. Sorption kinetics results for ^{125}I tracer on a set of minerals (Fuhrmann et al., 1998)

Endemic Goitre in Sri Lanka

Sri Lanka located close to the equator is a typical humid tropical country and is an ideal case study for medical geology. About 75% of the 20 million population of Sri Lanka live in rural areas and depend on the immediate physical environment for their food, water and other basic amenities. The geochemistry of soil and water therefore has a marked effect on the health of the population and significant correlations exist between certain diseases and the geochemistry of some elements (e.g. F and dental diseases).

Endemic goitre is a major national health problem in Sri Lanka and a salt iodization programme is now being carried out. It has been roughly estimated that nearly 10 million people are at risk from IDD. Endemic goitre has been reported in the wet zone of Southwest Sri Lanka for the past 50 years but rarely occurs in the more northern dry zone (Fig. 5.13). The IDD prevalence in the districts of Sri Lanka is shown in Table 5.9.

Early reports on endemic goitre include Wilson (1954), Mahadeva et al. (1968), Gembicki et al. (1973), Piyasena (1979) and Fernando et al. (1987, 1989). In these studies there was very little emphasis on the geochemical aspects of the endemicity of goitre. Dissanayake and Chandrajith (1993) considered for the first time, geochemical factors that may have a bearing on the prevalence of goitre in Sri Lanka. Table 5.10 shows some geochemical data on the water and soil in areas of Sri Lanka as studied by these authors. Cluster analysis of the geochemical data showed that the endemic goitre region lies in the group with lowest I, alkali earths, Cl^-, NO_3^-, Fe and Mn. Sri Lanka, being a small island, is expected to receive a higher iodine input from the sea and it is both interesting and surprising to note that the Kalutara district (located in a coastal area) has a 44% rate of endemic goitre. This study clearly showed the importance of geochemical goitrogens in the aetiology of IDD. Dissanayake and Chandrajith (1993) suggested that elements such as Co, Mn, Se, F, As, Zn, Ca, Mg, Cu and Mo may also have an influence in the aetiology of IDD.

Balasuriya et al. (1992) studied 609 sample of drinking water collected from scattered sources from the eight districts of Kandy, Matale, Kalutara, Anuradhapura, Polonnaruwa, Colombo, Puttalam and Gampaha for the iodine contents using the electrode method. Table 5.11 shows the data obtained and their rank orders.

Table 5.9. Prevalence of iodine deficiency disorders in different districts of Sri Lanka (Fernando et al., 1989)

	Male	Female	Both sexes	≥13-18 yrs.	N
Kalutara	39.7	49.5	44.9	24.1	13373
Kegalle	18.7	34.1	26.4	11.1	2549
Monaragala	18.5	31.0	25.0	11.1	1178
Kandy	15.4	27.8	22.2	8.0	7840
Ratnapura	14.0	28.3	20.5	5.3	2070
Hambantota	15.5	23.9	19.7	4.2	2474
Galle	11.0	21.9	17.9	8.8	1926
inBadulla	14.5	20.5	17.6	7.6	856
Puttalam	10.3	13.4	12.3	2.5	2625
Kurunegala	8.3	14.1	11.1	3.5	3445
Colombo	6.7	8.2	7.6	1.8	1373
Anuradhapura	4.6	10.1	7.3	1.1	1081
Matale	3.7	8.9	6.5	2.1	1513
Chilaw	4.1	8.1	6.3	1.0	1208

In general, these authors observed that the iodine levels in the water in Sri Lanka are higher than the levels reported from other countries and only 20% of the samples had values below 10 µg/L. Their important findings were:

- There is a geographic variation in the iodide content of drinking water.
- The iodide content is related to the depth of the source (deep wells had the higher iodide levels).
- The difference in iodide contents of drinking water in cases of goitre and controls is minimal.

The Spearman Rank order correlation between water iodide and goitre prevalence among school children by district was -0.64 indicating that only about 40% of the variability of goitre prevalence between districts can be explained in terms of water iodide content.

This study also confirms the view that there are other geochemical factors that need detailed study in relation to IDD in tropical countries. Fordyce et al. (2000a) pursued the search for these geochemical goitrogens in Sri Lanka further and studied the selenium and iodine content in soil, rice and drinking water in relation to endemic goitre in Sri Lanka. They observed that the soil iodine concentrations in the Sri Lankan environment are average to marginal compared to soils elsewhere. This does not support the long-held theory that considers the soil in the wet zone to be depleted in iodine while the dry zone soils are enriched in iodine.

Table 5.10. Geochemical data on the water and soil in the Angunawala-Daulagala area in Sri Lanka (*Sample size 11) (Dissanayake and Chandrajith, 1993)

Parameter	Water (n = 60)		
	minimum	maximum	mean
pH	5.85	8.2	7.7
Alkalinity (mg/L)	30	420	138
F (µg/L)	44	700	297
Cl (mg/L)	6	108	35
I (µg/L)	15	150	55
NO_3 (mg/L)	1.5	15	8.5
Na (mg/L)	27	1016	512
K (mg/L)	0.55	9.4	7.3
Ca (mg/L)	1.35	1616	50
Mg (mg/L)	0.23	16.64	5.9
Mn (µg/L)	1	208	53
Fe (µg/L)	520	2430	1166
Hardness (mg/L)	7	341	82
Co (µg/L)*	1	23	11
	Soil (n = 60)		
pH	3.8	6.8	5.16
F (mg/kg)	0.4	45	8.4
Cl (mg/kg)	8.0	432	155
I (mg/kg)	0.007	6.5	2.0

Table 5.11. Rank order of goitre prevalence, mean and median iodine content of water in different districts of Sri Lanka (Balasuriya et al., 1992)

District	Mean (µg/L)	Median (µg/L)	Rank order of median water iodine	Rank order of goitre prevalence
Kandy	30.96	19.1	5	7
Matale	16.91	11.1	2	5
Kalutara	15.50	12.2	4	8
Anuradhapura	118.03	101.6	8	1
Polonnaruwa	47.21	33.0	7	2
Colombo	16.86	11.4	3	4
Puttalam	34.68	21.6	6	3
Gampaha	11.90	5.0	1	6

Fordyce et al. (2000a) studied 15 villages with different levels of IDD (Figure 5.14) selected from both the dry and wet zones of Sri Lanka. The results showed that concentrations of soil total Se and I are highest in the HIDD (high goitre incidence) villages. The soil clay and organic matter appeared however, to inhibit the bioavailability of these elements.

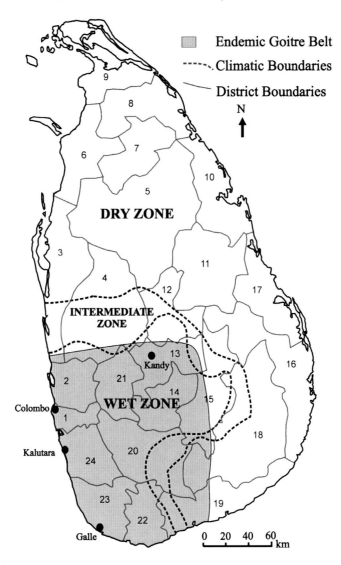

Fig. 5.13. Endemic goitre region of Sri Lanka (Districts:- 1- Colombo; 2- Gampaha; 3- Puttalam; 4- Kurunegala; 5- Anuradhapura; 6- Mannar; 7-Vavuniya; 8- Mullaitivu; 9- Jaffna; 10- Trincomalee; 11- Polonnaruwa; 12- Matale; 13- Kandy; 14- Nuwara Eliya; 15- Badulla; 16- Batticaloa; 17- Ampara; 18- Monaragala; 19- Hambantota; 20- Ratnapura; 21- Kegalle; 22- Matara; 23- Galle; 24- Kalutara)

Iodine Geochemistry and Health

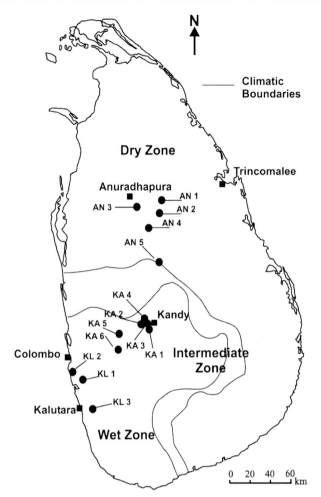

Fig. 5.14. Sketch map showing the location of the 15 study villages. NIDD- No/low goitre incidence (<25%); MIDD- Moderate goitre incidence (10-25%) and HIDD- High goitre incidence (>25%) (Modified after Fordyce et al., 2000a)

Concentrations of iodine in rice were low (≤58 ng/g) (Table 5.12) and rice is therefore not a significant source of iodine in the Sri Lankan diet. However the iodine levels in Sri Lankan rice was comparable to those in rice from other parts of the world. High concentrations of iodine (up to 84 µg/L) in drinking water in the dry zone was considered to be one of the sources of iodine. Although selenium-deficiency was not restricted to areas where goitre was prevalent, a combination of iodine- and Se-deficiency was considered to be involved in the pathogenesis of goitre in Sri Lanka.

Table 5.12. Summary of iodine and Se determinations in all sample types in each goitre incidence village group (Abbreviations: NIDD, no/low goitre incidence; MIDD, moderate goitre incidence; HIDD, high goitre incidence; nd, no data) (Fordyce et al., 2000a)

Group	Sample Type	Min. Se	Max. Se	Mean Se	No.	Min. I	Max. I	Mean I	No.
NIDD	Soil (ng/g)	113	663	226	25	130	10000	2260	25
	Rice (ng/g)	6.8	150	42	25	45	58	51	5
	Water(µg/L)	0.06	0.24	0.11	5	53	84	66.5	5
	Hair (ng/g)	104	765	294	25	nd	nd	nd	
MIDD	Soil (ng/g)	310	5238	875	24	130	6600	2008	25
	Rice (ng/g)	0.1	776	55	25	<38	<38	<38	5
	Water(µg/L)	0.06	0.09	0.07	5	3	23.5	5.5	5
	Hair (ng/g)	118	2652	389	25	nd	nd	nd	
HIDD	Soil (ng/g)	276	3947	1124	25	1000	9600	3914	25
	Rice (ng/g)	0.1	127	25	25	<38	<38	<38	5
	Water(µg/L)	0.06	0.09	0.07	5	3.3	20.2	7.02	5
	Hair (ng/g)	111	984	302	25	nd	nd	nd	

The Endemic Goitre Belt of India and Maldives

Among the nutritional disorders in India, IDD is widely prevalent with an estimated 120 million people affected (ICMR, 1989; Pandav and Kochupilai, 1982). The northern and northeastern parts of India have high endemic goitre incidence (range 25-54%) and these regions belong to the great arc of the Himalayas from West Pakistan across India and Nepal, into northern Thailand, Vietnam and Indonesia which forms one of the most highly endemic regions of the world.

Longvah and Deosthale (1998) analysed food and groundwater samples collected from the North Eastern states of Nagaland, Manipur, Mizoram, Meghalaya, Arunchal Pradesh, Assam and Sikkim for iodine. Figure 5.15 and Table 5.13 show the districts studied and their water-iodine contents. The mean iodine content in water samples from the goitre-endemic states of northeast India ranged from 6.65 µg/L in Sikkim to 8.89 µg/L in Assam. The iodine values groundwaters of northeast India ranged from 3.0 to 31.5 µg/L, and this was much lower than the range 30-50 µg/L observed in the non-endemic areas of Gujarat, Maharashtra, Mysore and Madhya Pradesh as reported by Tulpule (1969). The state of Sikkim, which had the lowest water- iodine content, has an incidence of goitre of 54% (Pulgar et al., 1992).

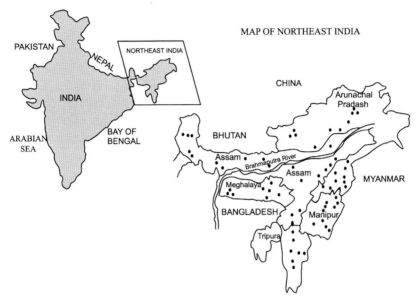

Fig. 5.15. Map of India and the location of groundwater sampling sites for iodine contents (Table 5.12) (Longvah and Deosthale, 1998)

In the case of food, the mean iodine content of rice from the goitre-endemic states of northeast India was, lowest in Sikkim (8.8 µg/100g) and highest in Assam (12.9 µg/100g). In non-endemic areas such as Hyderabad, the iodine in rice had an average of 40 µg/100µg. Other foods such as maize were also found to contain only about 25% of the iodine values observed in non-endemic areas.

Table 5.13. Iodine content in groundwater (µg/L) samples from northeast India (after Longvah and Deosthale, 1998) and Hyderabad (Mahesh, 1993)

State	No. of Samples	Range (µg/L)	Mean (µg/L)±SD
Arunachal Pradesh	24	3.2-14.5	6.98 ±2.1
Assam	44	4.3-31.5	8.89±4.9
Manipur	51	3.8-22.0	7.80±2.8
Meghalaya	35	4.3-14.7	7.68±1.84
Mizoram	48	4.0-13.9	6.92±1.7
Nagaland	54	3.2-11.9	6.57±1.5
Sikkim	31	3.0-12.6	6.65±1.8
Pooled data	287	3.0-31.5	7.38±2.7
Hyderabad	50	5.0-63.7	36.5±4.8

Even though the main endemic-goitre belt lies in the north and northeast of India, it has been found that based on results of sample surveys conducted by different agencies in 275 districts of 25 states and 4 union territories, 235 districts have been identified as endemic for IDD (Tiwari et al., 1998; ICMR, 1989).

The Kottayam district, in Kerala state, South of India, was studied for its iodine deficiency by Kapil et al. (2002). A total of 1872 children in the age group 6-12 years were studied and clinically examined. The total goitre prevalence was found to be 7.05% and this indicated that the population is in a transitional phase from iodine deficient, as related by the goitre rate, to iodine sufficiency, as revealed by the mean urinary iodine excretion level of 175 µg/L.

The importance of natural goitrogens in IDD is clearly seen in the case of Maldives, a country of islands. Being under the major influence of the sea, the main iodine reservoir, one expects a very low IDD rate in Maldives. Pandav et al. (1999), who studied IDD in the Maldives, observed that the total goitre rate was 23.6%, with grade I goitre contributing 22.5% to this figure.

Goitre in Vietnam

Vietnam is another country in South Asia which has been affected by iodine deficiency disorders. In 1992, Vietnam had 40 provinces with IDD at a rate of 43.2%. It was estimated that a million people were affected (Binh et al., 1992). As shown in Figure 5.16, the most affected areas were Ham Yen, Bach Thong, Cho Don, Sapa, Ngugen Binh and Cumngat. However, Vietnam has embarked on a successful salt iodisation programme and the IDD rate in Vietnam has dropped sharply.

Iodine Deficiency in China

In some parts of China, iodine deficiency disorders are common and have affected millions of people, bearing in mind that China is the most populous country of the world. Fordyce et al. (2002) studied the environmental controls in IDD in Xinjiang Province of China, known to be an area badly affected by IDD. According to the 1995 National IDD Survey, the goitre prevalence rate among school children aged 8 to 10 was 43.3%, the highest in China (Shaohua and De Long, 2000). This area has a semi-arid climate with an annual rainfall of less than 100 mm.

Iodine Geochemistry and Health

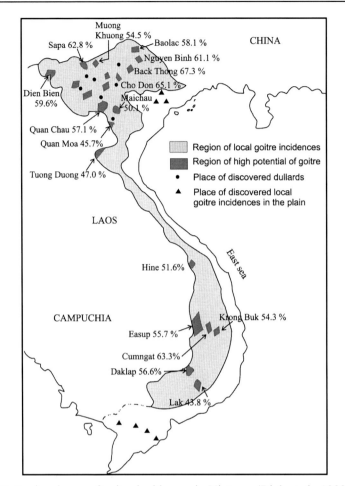

Fig. 5.16. Regional map of goitre incidences in Vietnam (Binh et al., 1992)

Some previous workers carried out a new project based on adding iodine (iodine dripping) to irrigation water and Fordyce et al. (2002) examined the environmental iodine and impact on health in three contrasting areas.

Area 1 - AC 148 low (3.5%) recent goitre prevalence (20% historic rate). no iodine irrigation. Iodized salt available

Area 2 - Kuqu District >30% goitre rate. No iodine irrigation. Iodized oil programme implemented.

Area 3 - Wushi District 40-60% goitre rates. Iodine irrigation. iodized oil programme implemented.

Table 5.14 shows the village median iodine contents in soil, water, wheat and cabbage samples from the three areas studied. The results from 5 soil and wheat, 3 cabbage and 1 drinking water samples collected in 5 villages in each of the three study areas showed that the iodine concentrations in soils are similar in all three areas, and low by world standards. The total iodine content of wheat in the three study areas is broadly similar and comparable to other areas of the world. Cabbage iodine contents also showed little variation in the three study areas, but are slightly lower than results from other areas.

Table 5.14. Village, median iodine contents in soil, water, wheat and cabbage samples (No. = No of Samples; WSol = Water Soluble; DW = Dry Weight) (Fordyce et al., 2002)

Area	Village	Soil Total I, mg/kg	Soil WSol. I mg/kg	Wheat I µg/kg DW	Cabbage I µg/kg DW	Water I µg/L
148	Village 2	1.17	0.012	107.0	106.6	92.25
148	Village 21	1.14	0.012	171.4	103.5	100.00
148	Village 3	1.16	0.020	105.9	118.6	93.00
148	Village 4	0.67	0.020	182.0	94.9	80.00
148	Village 6	0.77	0.012	140.6	155.3	78.00
Kuqa	Qiman	0.99	0.020	132.2	71.7	3.09
Kuqa	Sandaqiao	0.93	0.012	101.2	77.6	3.15
Kuqa	Waqiao	1.16	0.012	80.3	118.4	4.05
Kuqa	Wuzun	1.00	0.012	112.6	178.6	2.40
Kuqa	Yaha	1.65	0.012	162.9	154.5	3.25
Wushi	Aheya	1.37	0.012	156.8	93.5	0.10
Wushi	Autebeixi	0.70	0.012	111.4	106.6	0.40
Wushi	Daqiao	0.62	0.012	121.4	79.6	3.70
Wushi	Wushi Town	0.94	0.014	152.6	81.7	0.95
Wushi	Yimamu	0.98	0.012	121.1	176.9	2.10

It is known that acidic soils favour and alkaline soils inhibit the bioavailability of iodine. The very low iodine status of the study area was explained by their location in an alkaline desert environment.

A feature worthy of note in this study was that the surface and shallow groundwaters used for drinking in Kuqa and Wushi Districts had very low iodine contents (0.1 to 4.05 µg/L) whereas waters taken from deep bore holes in the AC-148 area contained very high quantities of iodine (78-100 µg/L). Drinking water was therefore an important source of iodine in this case. Fordyce et al. (2002) were therefore of the view that the eradication of goitre from the AC 148 region many have been due to development of the area and provision of centralized groundwater supplies, rather than to the use of iodized salt. In subsistence populations consuming low-iodine food, water from iodine-rich deep sources can therefore be an important dietary source.

Iodine Deficiency in East Africa

Many countries of Africa, notably those lying in and around the tropical belt are particularly vulnerable to several diseases linked to nutritional deficiency. The non-availability of essential trace elements, poor diet, poverty, bad sanitary conditions among others, cause a multitude of health-related problems for a large population. Among these, IDD rank very high.

In 1993, Africa ranks third among regions most affected by IDD, after Western Pacific and Southeast Asia. About 220 million people were estimated to be at risk of whom 95 million were goitrous. Each year, about 3 million women who were "at risk" became pregnant resulting in about 15,000 foetal deaths, the birth of 30000 cretins and about 1 million brain-damaged children (WHO/MDIS-1993). At present however, due to salt iodization programmes, this figure has dropped dramatically.

As shown in Figure 5.17 several eastern African countries have high endemic goitre prevalence. In Kenya, goitre rates varying from 15 to 72% have been observed (Davies, 1996) and the highest rates being noted in the Highlands of the Rift Valley, Central Nyanza and western Province. Two possible goitrogens that may have contributed to the low iodine status of Kenya were the complex ions thiocyanate (SCN^-) and fluoroborate (BF_4^-). The BF_4^- ion was considered to be present in natural waters in parts of the Rift Valley (Davies, 1994). The univalent BF_4^- ion is known to have the same ionic size as the iodide ion and its goitrogenic properties have been reported (Langer and Greer, 1977).

In Uganda IDD, mainly goitre was a major surgical problem in the national hospitals in the 1960s (Kajubi, 1971). In 1991, a local survey was carried out in four mountainous districts and five lower altitude districts reported a high total goitre rate among school children 6 to 12 years old and which ranged from 63 to 76%. The mountainous districts had higher goitre rates (Bachou, 2002). In 1999, as a result of a successful salt iodisation programme there was a significant reduction in the overall total goitre rate to 16%.

Zaire is another tropical country that has been severely affected by IDD and a large scale eradication programme was required. Several regions of northern Zaire have been documented as areas with severe endemic goitre (Thilly, 1992). About 4 million people were affected and the rate of goitre exceeded 60% in adult women and cretinism was diagnosed at 1-10% of the population. The goitrogen role of thiocyanate in food containing cassava has been established.

Ngo et al. (1997) studied the selenium status in pregnant women of a rural population in Zaire and its relationship to iodine deficiency. They determined serum selenium and thyroid function parameters including urinary iodide in 30 prenatal clinics of rural villages distributed throughout the country. In all cases except one, biochemical maternal hypothyroidism (serum TSH >5 mIU/L) was present, with a frequency ranging from 3% to 12%. Hypothyroidism is caused by insufficient production of thyroid hormones by the thyroid gland.

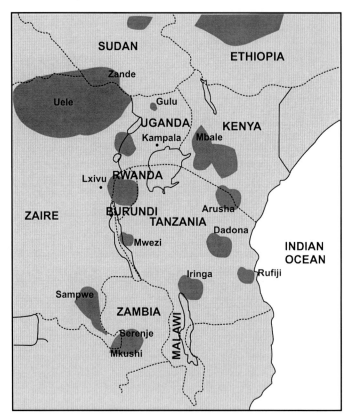

Fig. 5.17. Distribution of reported goitre areas in Eastern Africa (Davies, 1996)

In Nigeria, IDD was a major health concern about 3 decades ago (Ekpechi, 1967). The total goitre rate for Nigeria was placed at 20%. Nigeria has a well-demarcated goitre belt where almost all the inhabitants within the belt live on cassava–based food. At least 60 million Nigerians (from a population of about 120 million) were at risk of IDD. This rate has now dropped and Nigeria appears to have gained success in IDD eradication.

In 1988, a National Goitre Survey carried out in Zimbabwe highlighted goitre as a major health problem (WHO/MDIS, 1993). The National Visible Goitre Rate (VGR) was 3.7% while the Total Goitre Rate (TGR) reached 42.3%. From the 53 districts surveyed, 20 had severe endemia and 21 moderate endemia. Most of the regions with severe endemia were in the north-eastern mountainous region with a very high rainfall. By 1966, however, after a successful salt iodization programme, IDD was virtually eliminated. However, thyrotoxicosis, caused by excess intake of iodine had been recorded.

CHAPTER 6

NITRATES IN THE GEOCHEMICAL ENVIRONMENT

Nitrates in the geochemical environment, particularly in the aquatic environment, are of special importance in view of health implications. With great emphasis being placed on increased food production and the concomitant use of nitrogenous fertilizers, nitrates figure prominently in the medical geochemistry of cancer and some other diseases notably in developing countries. Among the major threats to groundwater from which drinking water supplies are obtained are leachates from human and animal waste matter. Additionally industrial and agriculture leachates mostly from large farmlands contribute quite significantly to groundwater pollution.

The source of nitrates, their pathways in the geochemical cycle across the soil-water-plant and human/animal systems and their impact on the latter constitute an interesting study within the field of medical geology.

THE NITROGEN CYCLE

In the terrestrial ecosystem, the nitrogen cycle represents one of the most important nutrient cycles (Chapin et al., 2002; Smil, 2000; Townsend, 2007; Sprent, 1987). As an essential component of proteins, nucleic acids and other cell constituents, nitrogen plays a major role in plant physiology and is required in substantial quantities. Even though there is an abundant source of nitrogen in the earth's atmosphere, much of it, ~79% is in the form of a nearly inert gas and is hence not available to most organisms. For plants to use this nitrogen for their growth, it should be in the form of NH_4^+ and NO_3^- ions. The former is used less by plants for uptake as it could in large concentrations be toxic. Nitrates therefore are the most important form of nitrogen carriers in the growth cycle of plants. Figures 6.1 and 6.2 illustrate the nitrogen cycle in nature and the nitrogen transformations in soil.

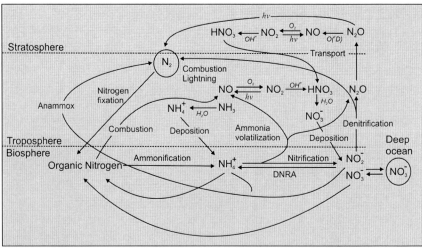

Fig. 6.1. The nitrogen cycle [Chemical species over arrows indicate reaction with that species to effect the transformation. Fixation, ammonification, assimilation, anammox, dissimilatory nitrate reduction to ammonia (DNRA), nitrification, and denitrification are enzyme-catalyzed processes-(graphics by Darrouzet-Nardi, 2005)]

While most plants obtain their nitrogen as inorganic nitrate from soil solutions in most ecosystems, nitrogen is stored primarily in living and dead organic matter. The transformation of organic nitrogen into inorganic nitrogen in the upper soil layers, is brought about by decomposition caused by the action of microorganisms. These are carried out by some types of bacteria, actinomycetes and fungi. Some bacteria convert N_2 into ammonia by the process of nitrogen fixation and these may be free-living or living in symbiotic associations with plants or other organisms. Ammonium is sorbed on surfaces of clay particles in soils due to the positive charge and they are held by soil colloids. When the ammonium ion is released from the colloid surface by cation exchange, bacteria belonging to the genus Nitrosomonas convert the ammonium ion into nitrite (NO_2^-) and another bacterium belonging to the genus Nitrobacter converts the nitrite to the nitrate (NO_3^-). These two processes, termed nitrification, involves oxidation.

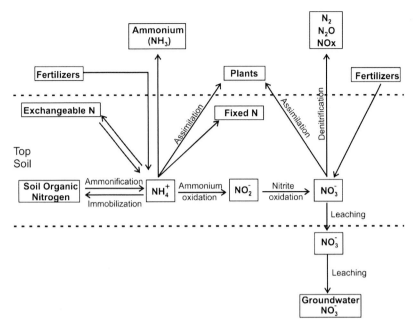

Fig. 6.2. Nitrogen transformations in soil (Rao and Puttanna, 2000)

Due to the very easy solubility of nitrates in water, leaching of nitrates from soils takes place rapidly, and these enter the hydrosphere through rivers and oceans. The process of denitrification, which is common in anaerobic soils, is carried out by heterotrophic bacteria and involves reduction of nitrates into the atmosphere. The oxygen needed for respiration by the bacteria is supplied by this process.

The nitrogen cycle as depicted in the Figures 6.1 and 6.2 is quite often interrupted or altered due to anthropogenic activities. Nitrate leaching and rates of denitrification increase when nitrogen fertilizers are applied to vast tracts of land. Nitrate pollution and eutrophication of aqueous systems then occur with the enhanced growth of algae. Human and animal wastes add a considerable load to the nitrogen and ammonia budget in the geochemical cycle and these bring about an increased nitrate concentration in the ground and surface waters.

The fact that amino acids which produce proteins are nitrogen-rich makes nitrogen an essential element. One of the most interesting features of nitrogen on earth therefore is the effect of its availability on living organisms and the ecosystems in which they live. Remote sensing of the earth's nitrogen cycle is assuming greater importance in quantifying biogeochemical cycles (Darrouzet-Nardi, 2005). As shown by Galloway et al. (2003, 2004) and Vitousek et al. (1997), the anthropogenic impact and alteration of the cycle is of much greater significance than ever before. The fixing of nitrogen by humans for use as fertilizer in agriculture and as a by product of fossil fuel combustion, though of great benefit in agriculture and transportation also has some negative influences (Darrouzet-Nardi, 2005). These may appear in the form of changes in the ecosystems manifesting further as changes in the physiological characteristics of organisms (Bowmann, 2000; Aber et al., 1989). Apart from the biological effects of fertilizer, reference has also been made to:

(i) public health concerns
        (a) as unbalanced diets
        (b) respiratory ailments
        (c) cardiac diseases
        (d) cancer
        (e) allergic pollen production
        (f) dynamics of some vector-borne diseases (Townsend et al., 2007);

(ii) ecosystem acidification as nitric acid in precipitation (Schindler et al., 1985; Aber et al., 1989; van Breeman et al., 1982)

(iii) global warming as nitrous oxide (Houghton, 2001).

NITRATES, FERTILIZERS AND ENVIRONMENT

In order to feed the expanding world population, science-based agriculture has relied heavily on fertilizers, to increase crop production. In the tropics, where intense leaching of soils takes place, nutrients are rapidly lost and fertilizers are used extensively to replenish the essential nutrients N, P, and K.

Developing countries have around 76% of the global population and N-fertilizers are applied in quantities greater than that in developed countries. Figure 6.3 illustrates the world fertilizer consumption and Figure 6.4 shows the fertilizer consumption in South Asia, where India accounts for nearly 80% of fertilizer consumption. Pakistan and Bangladesh also have large fertilizer requirements.

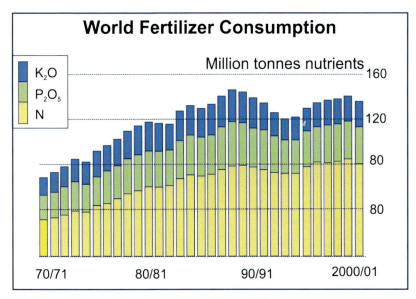

Fig. 6.3. Fertilizer nutrient consumption (IFA, 2004)

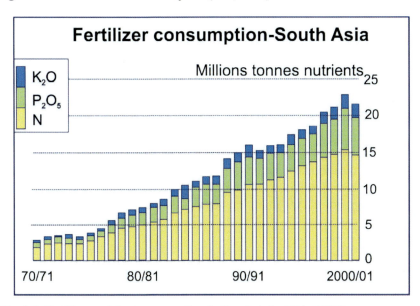

Fig. 6.4. Fertilizer nutrient consumption in South Asia (IFA, 2004)

Global fertilizer use increased at an annual rate of 5.5% from 27.4 million nutrient ions in 1959/60 to 143.6 million tons in 2001. In all developing regions, fertilizer use increased significantly (Table 6.1). The use of N-fertilizers by far outweighed that of phosphate and potash. In 1994/95,

N-fertilizers accounted for 64% of the fertilizers consumed by developing countries, as compared to 25% for phosphate and 11% for potash.

Table 6.1. World Fertilizer Use (IFA, 2004) Notes: East Asia excludes Japan. West Asia/North Africa excludes Israel

Region/Nutrient	Fertilizer Use			Annual Growth	
	1959/60	1980/90	2020	1960-90	1990-2020
	(million nutrient tons)			(percent)	
Developed countries	24.7	81.3	86.4	4.0	0.2
Developing countries	2.7	62.3	121.6	10.5	2.2
East Asia	1.2	31.4	55.7	10.9	1.9
South Asia	0.4	14.8	33.8	12.0	2.8
West Asia/North Africa	0.3	6.7	11.7	10.4	1.9
Latin America	0.7	8.2	16.2	8.2	2.3
Sub-Saharan Africa	0.1	1.2	4.2	8.3	3.3
World total	27.4	143.6	208.0	5.5	1.2
Nitrogen	9.5	79.2	115.3	7.1	1.3
Phosphate	9.7	37.5	56.0	4.5	1.3
Potash	8.1	26.9	36.7	4.0	1.0

With the increasing use of N-fertilizers, mostly in developing countries, the possibility of nitrate pollution of groundwater will be strongly associated with fertilizer-N use efficiency (Singh et al., 1995). These authors have pointed out that due to the poor fertilizer-N use in many developing countries, notably in the irrigated soils of Asia and humid tropics of Africa, the potential for nitrate pollution of groundwater is quite large. Intense irrigation, apart from high rainfall, is now available to farmers of Asia and the increase of N-fertilizer use can be viewed as a future environmental hazard affecting the groundwater systems. It should be bear in mind that three quarters of the world population live in developing countries and a majority (60%) are engaged in farming. Farmers in developing countries possess ~54% of the arable land available in the world (FAO, 1991).

The proper fertilizer use is an important pre-requisite for retarding nitrate pollution of groundwater. The percent recovery of fertilizer-N by a crop often termed "fertilizer use efficiency" varies with the N source and the rate at which fertilizer is applied, the nature of the chemical and biochemical reactions between soil and fertilizer, the timing and placement of the fertilizer, the nature of crop and its N-requirement, the adequacy of other nutrients and a host of soil, climatic and management factors (Singh et al., 1995). When conditions are very favourable, ≥80% of the fertilizer-N may be recovered by the crop, but under most situations, efficiencies of ≤50% are common (Allison, 1966).

The excess accumulation of nitrogen in soil in relation to crop yield and N-fertilizer application is shown in Figure 6.5. In such cases where there is a large amount of residual nitrogen in soil, the potential for nitrate pollution is significantly high. In the highly permeable Ultisols and Oxisols in the humid regions where precipitation greatly exceeds evapotranspiration during the growing season, there can be extensive N-leaching. Even ≥50% of mineral-N initially in the soil may be lost through leaching between the onset of the rain and plant establishment (Osiname et al., 1983).

Leutenegger (1956) reports that in Tanzania only 7% of the fertilizer-N was recovered within the 120 cm depth of the uncropped bare plot one year after application. In Nigeria, in the humid tropics, both heavy rainfall and irrigation is present. It has been estimated (Adetunji, 1993) that ~30% of the N-applied to maize was lost below the root zone and over a 3 year period, the nitrate pollution of groundwater exceeded the maximum level accepted for potable water by a significant margin. In the city of Sokoto, Nigeria, Uma (1993) observed that in waters of over 40 wells in a shallow lateritic aquifer nitrate enrichment ranged from 20 to 100 mg/L.

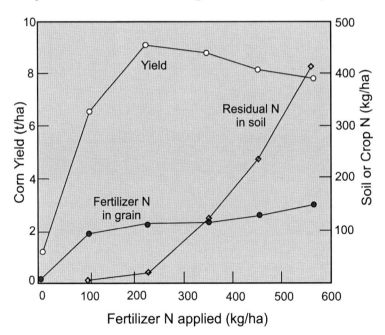

Fig. 6.5. Fertilizer application and crop yield (Broadbent and Rauschkolb, 1977)

Figure 6.6 illustrates a case study in Sri Lanka where the nitrate levels are exceptionally high due to excessive application of nitrogenous fertilizer. In view of the fact that the vast majority of the people living in developing countries of the tropics obtain their water supplies directly from the ground, the levels of nitrate fertilizers in groundwater are of vital importance.

There are several N-fertilizers that are available. Among these are: (a) anhydrous ammonia (82% N)- a liquid under high pressure (b) mixture of urea and ammonium nitrate (28 - 32% N) (c) aqua ammonia (21% N)- liquid under low pressure (d) urea (46% N) (e) ammonium nitrate (33% N) (f) ammonium sulphate (21% N) (g) calcium nitrate (16% N) (h) potassium nitrate (13% N) (i) sodium nitrate (16% N). From among these, urea [CO(NH$_2$)$_2$] is a widely used dry N-fertilizer. Upon application to the soil, urea is converted to ammonia which reacts with water to form ammonium within two to three days. The NH$_4^+$ ion then gets converted to nitrate (NO$_3$) through nitrification.

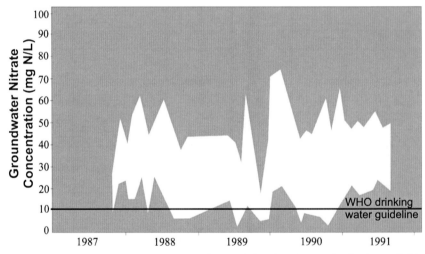

Fig. 6.6. Envelope of nitrate concentration in irrigation dug wells in the Kalpitiya Peninsula, Sri Lanka (source British Geological Survey, Technical Report WD/OS/86/21; reproduced with kind permission from the British Geological Survey)

NITROGEN LOADING IN RICE FIELDS

In South and Southeast Asia, rice (*Oryza sativa*) is the most important food, and is the only crop grown under flooded soils. The global rice area is 150.7 million ha; more than 90% of which lies in Asia (FAI, 1997). India is known to have the worlds largest rice cultivation area (47.2 million ha), followed by China (31.3 million ha) (Ghosh and Bhat, 1998). Rice cultivation uses a large share of fertilizer; urea being the most commonly used nitrogen-fertilizer. Since for most of the growing season rice fields are submerged, nitrate losses from rice fields can be high. Prediction of nitrate losses is complicated by the variable nature of soils and the complex set of N-transformation processes taking place under flooded soil conditions (Chowdary et al., 2004).

During submergence of the soils of the rice fields, changes in physico-chemical and biological properties of the soils take place. The root zone of the rice plant gets converted from an aerobic to an anaerobic environment, the process being governed mostly by microorganisms. As shown in Figure 6.7, the rice plant obtains its nitrogen fixation by algae and bacteria of inherent NH_4^+ and NO_3^- nitrogen in the soil and addition of nitrogen through chemical fertilizers (Ghosh and Bhat, 1998).

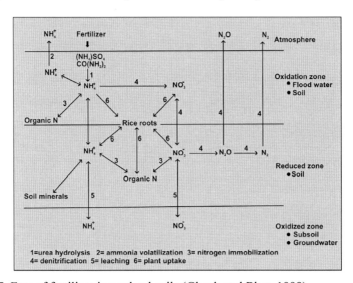

Fig. 6.7. Fate of fertilizer in wetland soils (Ghosh and Bhat, 1998)

Part of the nitrogen fertilizer applied to the rice field undergoes leaching and percolates down the soil profiles and mixes with the groundwater, which then gets contaminated with NO_3-N. In India, high nitrate concentrations have been reported in some states such as Maharashtra, Tamilnadu, Haryana, Rajastan, and Andra Pradesh (Ozha et al., 1993; Vijay Kumar et al., 1993) as well as in areas in Punjab and Delhi (Rao and Puttanna, 2000). It was observed that 70% of the wells surveyed contained nitrate exceeding the permissible limit of 45 mg/L (=10 mgN/L). A level as high as 7400 mg/L indicating extreme contamination had also been recorded.

NITRATES FROM HUMAN AND ANIMAL WASTES

The risks to health caused by groundwater contamination from on-site sanitation are a concern in many overpopulated developing countries notably in the tropics. On-site sanitation systems dispose of human excreta into the ground, mostly from pit latrines and septic tanks. Sanitation has been defined as *"the means of collecting and disposing of excreta and community liquid wastes in a hygienic way so as not to endanger the health of individuals and the community as a whole"* (WHO, 1987).

The two main health risks associated with the degradation of water quality due to on-site waste disposal systems are (a) faecal-oral disease transmission and (b) nitrate poisoning. Pollution from on-site waste disposal is influenced by a variety of complex factors (Fourie and van Ryneveld, 1995):

i) Varying sub-surface conditions: in addition to the variety of sub-surface soils present, within any soil the most critical distinction is between the saturated and the unsaturated zone.

ii) Varying contaminants: different contaminants show different characteristics, such as mobility and persistence, and these are affected in different ways in the sub-surface.

iii) Varying polluting mechanism and movement of pollutants.

Pathogens are defined as disease-causing organisms. Human excreta are known to contain worm eggs, protozoa, bacteria, and viruses and from among these, eggs and protozoa are effectively screened by soil during groundwater flow. The smaller bacteria and viruses are therefore of the greatest health concerns, bearing in mind that in most developing countries, water-borne diseases affect millions of people. Pathogens cannot travel farther and faster than the water in which they are suspended and groundwater hydrology therefore has a marked influence on aquifer pollution. As noted by Cave and Kolsky (1999) the key factor that affects the removal and elimination of bacteria and viruses from groundwater is the maximisation of the effluent residence time between the source of contamination and the point of water abstraction. Because of the very low velocities of unsaturated flow, the unsaturated zone is the most important line of defence against faecal pollution of aquifers.

From a geochemical standpoint, nitrates and their pathways in the aqueous systems closely associated with on-site waste disposal, have aroused great interest. Nitrates are considered as significant health indicators:

(a)   Excessive concentrations of nitrates are directly associated with methaemoglobinaemia or "blue baby syndrome"

(b)   Nitrates and nitrites are precursors of carcinogenic nitroso-compounds

(c)   Nitrates are useful as rough indicators of faecal pollution when microbiological data are lacking.

In the developing countries of the humid tropics, pit latrine soakaways (Fig. 6.8) often pose a threat to potable groundwater supplies in view of the large number of inhabitants confined to a smaller area and using water from wells located close to pit latrines.

Lawrence (1986) outlined the conditions under which water-supply tube wells in basement aquifers are most vulnerable to this type of pollution:

(i)   the water-table is shallow and the unsaturated zone is thin.

(ii)   the overburden is thin and pit latrines penetrate to, or close to, the top of the bedrock.

(iii)   groundwater movement is completely restricted to joints and fissures.

(iv)   population density is high (particularly urban fringe developments) where there is pressure to reduce the separation between latrines and water-supply wells.

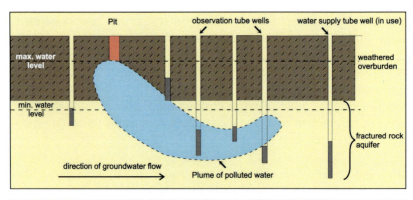
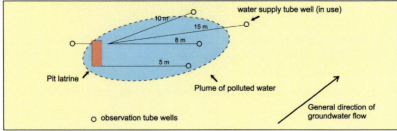

Fig. 6.8. Nitrate pollution by pit latrine soakaways in Sri Lanka. The bottom diagram illustrates the sketch plan of the site (Lawrence, 1986; reproduced with kind permission from the British Geological Survey)

(v) water-flush pit latrines are commonly used and this type of latrine increases the fluid loading and hence the likelihood of microbiological contamination of groundwater.

(vi) water-table gradients are steep, thus increasing the rate of groundwater movement

Figure 6.9 illustrates a case study in Sri Lanka showing the plume of polluted water from pit latrines in limestone terrain. Such cases are very common in the tropics and as outlined above, the general soil and climatic conditions aid in the mobility of the pollution species.

In Francistown located in Botswana, in mid 2000 nitrate pollution had reached alarming levels (Fig. 6.10). Areas with pit latrines, sewage ponds and/or cemeteries posed a groundwater hazard in terms of organic substances, bacteria and nitrates. In fact, because of the extreme nitrate pollution, nitrates served as the key indicator of overall groundwater quality in all environmental hydrogeology studies carried out by the Environmental Geology Division of BGR, Germany (Vogel, 2002).

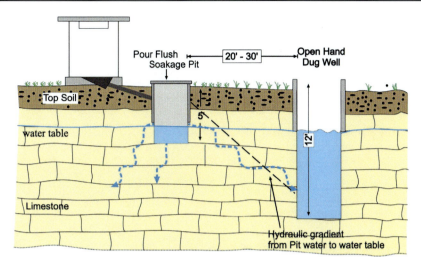

Fig. 6.9. Pit latrine in a limestone in Jaffna, Sri Lanka

Among the three main pollution sources namely (a) pit latrines (b) mine waste dumps (c) landfills, pit latrines were found to have had the worst impact on groundwater quality. The chemical analysis of water from a total of 48 public and private wells sampled within and around Francistown showed that nitrate concentrations were frequently well above the WHO maximum levels. In some cases, they had even reached levels between 100 to 300 mg/L (Mafa, 2003). By mid 2000, the major groundwater pollutant was nitrate where there was a clear correlation of elevated nitrate levels with pit latrines.

In Senegal in the Yeumbeul area close to Dakar, the capital city, high nitrate levels above the WHO limits have been observed (Tandia et al., 1999). In this area, shallow groundwater provided nearly all of the water supply for 7000 families from traditional wells. Close to every well is a family latrine and the distance separating the wells from the latrines varied between 2 and 36 m. The population had virtually no sanitation system and human waste was disposed directly into the environment causing a water quality hazard.

In the Dodoma area, located in a terrain of crystalline basement rock in Tanzania, in a semi-arid climate, nitrate levels in the groundwater had reached an average of 150 mg/L (Nkotagu, 1996). These high nitrate concentrations were found in both deep and shallow groundwaters and originated from sewage effluents.

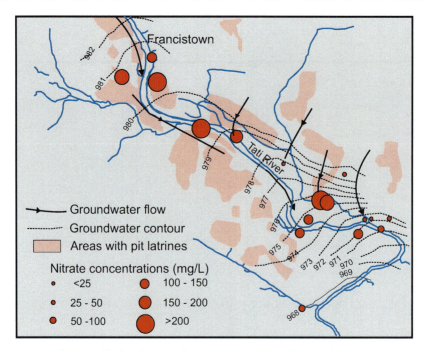

Fig. 6.10. Nitrate pollution in Francistown, Botswana (after Mafa, 2003)

In a different geological setting in the Essaouira Basin of Morocco, where fractured karstic materials were dominant, nitrate levels in groundwater had reached as much as 400 mg/L. The aquifer systems are interconnected through a network of fractures and stratification joints facilitating the transport of nitrogen loaded water (Laftouhi et al., 2003). Livestock excreta, agricultural fertilizers and human wastes were considered to be the sources of the excess nitrates.

A well water survey carried out in two districts of the Republic of Guinea (Gelinas et al., 1996) showed that there was widespread well water contamination from nitrate and faecal bacteria in both districts. In these areas, the water distribution network was primitive, domestic sewage untreated and sanitation only by poorly managed pit latrines and septic tanks. There was no storm water drainage system. As no fertilizers were used in the area, the very high nitrate levels were attributed to human excreta, domestic sewage aided by the high permeability of the soil. The high nitrate levels were correlated with high contents of organic matter probably due to incomplete degradation of organic nitrogen due to naturally low dissolved oxygen levels.

NITRATES AND HEALTH

Nitrates are natural components of all fruits and vegetables and as discussed earlier drinking water may also contain nitrates. They are also added to some meat and fish products as preservatives. Over the last few decades, dietary nitrate has been implicated in the formation of infantile methaemoglobinaemia and gastric/intestinal cancer through the production of carcinogenic N-nitrosamines. Comly (1945) first noted that consumption of drinking water environmentally contaminated with nitrate was associated with methaemoglobinaemia. The National Academy of Science of USA (1977) set a limit of 10 mgN/L as nitrogen (45 mg/L nitrate) as the maximum permissible level.

Nitrates and Methaemoglobinaemia

Nitrate itself is considered to be relatively non-toxic and it is the reduced species, nitrite that greatly increases the toxicity. Nitrate is reduced to nitrite by bacteria in the upper gastro-intestinal tract. Nitrite is then absorbed into the bloodstream, where it reacts with haemoglobin to form methaemoglobin (Bruning-Fann and Kaneene, 1993a). Methaemoglobin does not have the capacity to carry oxygen and this results in cyanosis and anoxemia if the level of methaemoglobin (met-Hb) reaches high levels. Cyanosis becomes evident when the concentration of met-Hb reaches about 10 g/L (5-10% of the total haemoglobin (Hb) is in the met-Hb form), becomes severe at 30 g/L and may result in death when levels exceed 60 g/L (Robertson and Riddell, 1949). It is the reduced oxygen delivery to the tissues that causes the death of the individual. Met-Hb however, is known to be converted back to Hb by the enzyme methaemoglobin reductace present in the erythrocytes. This enzyme normally keeps the met-Hb level between 1-2% but it is much less active in children under 3 months and hence the common occurrence of methaemoglobinaemia in such children.

In the case of animals, clinical signs of acute nitrate toxicity vary according to species. Nitrate is capable of inducing methaemoglobinaemia in a wide range of species e.g. cattle, sheep, swine, dogs, guinea pigs, rats, chicken and turkey. In general, ruminant animals develop methaemoglobinaemia while monogastric animals exhibit severe gastritis (Bruning-Fann and Kaneene, 1993b).

Nitrates and Cancer

Cancer, after heart disease is recognized as a leading killer disease. A large number of causative factors which have been isolated are in one way or another environmental. The relationship between cancer and the environment has been known since 1775, when a correlation between scrotal skin cancer and heavy exposure to soot among chimney sweepers was observed (Miller and Miller, 1972). Since then, many environmental pollutants have been shown to produce cancer in various parts of the body. Whereas the carcinogenicity of many laboratory synthesized chemical components have been the subject of intensive study, the effect of the natural environment on cancer is much less known. Epidemiological studies have indicated the importance of, for e.g. the quality of potable waters, the chemistry of the soils and the plants growing on them in geographically separated areas, quality of the air we breathe on human cancer. Since correlations by themselves rarely justify mechanistic interferences and only are recognised as weak suggestions of causality (Tannenbaum and Correa, 1985), evidence from epidemiology cannot often be used to draw conclusions on biological mechanisms.

However, epidemiology has its own virtues and if used intelligently, useful preliminary information may be obtained. There are certain parts of the world where some specific diseases, including cancer, show anomalous incidences clearly pointing to some features unique to that environment. Oesophageal cancer, for example, was highly prevalent in parts of Transkei (Laker et al., 1980), Iran, Central Asia and northern China (Kibblewhite, 1982). In these regions cancer was a more important cause of death than coronary heart diseases in Europe and USA. During the last 50 years, substantial evidence has been accumulated which show a relationship between geology, soils and climate and the widespread occurrence of oesophageal cancer. As pointed out by Laker (1979), such an integration of environmental factors leads to distinct soil properties and it has been proved possible to identify certain soil types as being common to areas of high cancer incidence. Figure 6.11 illustrates the relationship between environmental factors and incidence of cancer.

Nitrates in the Geochemical Environment

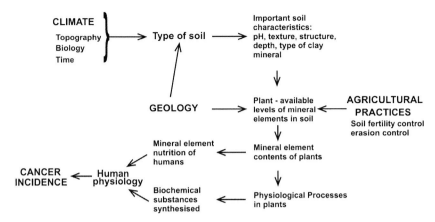

Fig. 6.11. Relationship between environmental factors and cancer incidence.

Nitrates can react with amines in the stomach or lungs to form N-nitrosamines, which have induced tumours in laboratory animals. Although the causation of human tumours is not directly linked to these compounds, exposure to them is considered as being potentially capable of initiating human cancer (National Research Council, 1978). Figure 6.12 illustrates the geochemical environment and the cancer forming nitrogen-bearing compounds.

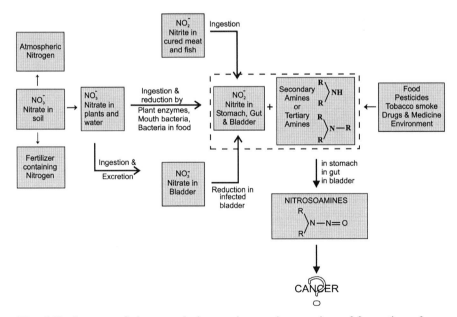

Fig. 6.12. Sources of nitrate and nitrosamines and some sites of formation of carcinogenic nitrosamines (Fishbein, 1979)

CHAPTER 7

MEDICAL GEOLOGY OF ARSENIC

INTRODUCTION

One of the worlds biggest environmental disasters, the arsenic poisoning of millions of people in Bangladesh and West Bengal, had its origins in the arsenic-bearing clay or peat layers at shallow depths within the Quaternary deltaic sediments of the Ganges-Brahamputra delta. As Berger (1999) pointed out, the irony was that the problem arose only recently as people started using water from tube wells drilled through international assistance programmes to provide cleaner water than available on the surface.

The magnitude of the problem of arsenic poisoning was so great that WHO having recognized the enormous health implications lowered the provisional guideline value for arsenic in drinking water from 50 µg/L to 10 µg/L. Among the countries which have well documented case studies of arsenic poisoning are Bangladesh, India (West Bengal), Vietnam, Taiwan, China, Mexico, Chile and many parts of the USA and Argentina (Figure 7.1). Recently, countries such as Nepal, Myanmar and Cambodia have also been added to the list of countries which have concentrations of As indicating water exceeding 50 µg/L in some sources. Table 7.1 summarizes the documented cases of naturally occurring As-poisoning in world groundwaters. A technical report of the British Geological Survey (2001) highlights the magnitude of the issue of arsenic poisoning in Bangladesh as follows.

"A survey of well waters (n =3534) from throughout Bangladesh, excluding the Chittagong Hill Tracts has shown that water from 27% of the shallow tube wells, that is wells less than 150 m deep, exceeded the Bangladesh standard for arsenic in drinking water (50 µg/L), 46% percent exceeded the WHO guideline value of 10 µg/L. Figures for 'deep wells' (greater than 150 m deep) were 1% and 5%, respectively. Since it is

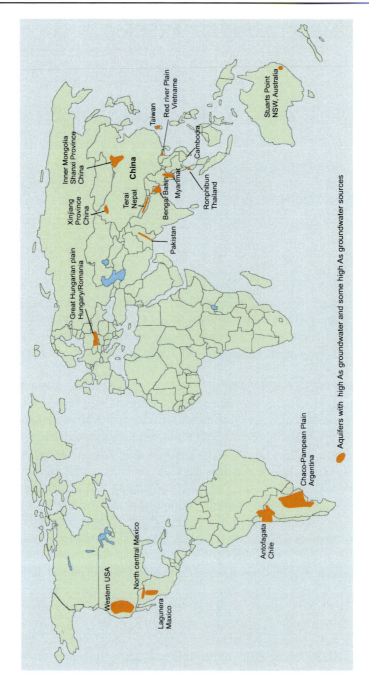

Fig. 7.1. Arsenic contaminated aquifers in the world (sources: Smedley and Kinniburgh, 2002)

believed that there are a total of some 6-11 million tube wells in Bangladesh mostly exploiting the depth range 10-50 m, some 1.5-2.5 million wells are estimated to be contaminated with arsenic according to the Bangladesh standard. Thirty five million people are believed to be exposed to an arsenic concentration in drinking water exceeding 50 µg/L and 57 million people exposed to a concentration exceeding 10 µg/L".

The above description clearly illustrates the enormity of the disaster and the importance of the understanding of the medical geochemistry of trace elements known to have an impact on the health of populations.

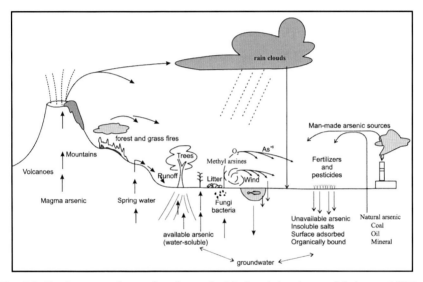

Fig. 7.2. Environmental transfer of arsenic (National Academy of Science, 1977)

The element arsenic (valence configurations $3d^{10}\ 4s^2\ 4p^3$) is a pollutant known to exist in four major oxidation states +5, +3, 0 and -3. It is found as a commonly distributed element in the atmosphere, rocks, minerals, soil, water and in the biosphere. Figure 7.2 is a diagrammatic representation of the arsenic cycle in nature. Arsenic mobilization in the environment occurs mostly under geogenic conditions though anthropogenic influences also affect the arsenic cycle to a significant degree. From among these, mining activity, combustion of fossil fuels, use of arsenic in agrochemicals is of particular importance. The medical geochemistry of arsenic is predominantly the aqueous chemistry of arsenic in the geological environment and its bioavailability. As shown in Figure 7.2, arsenic may be derived from a variety of sources and may enter the food chain through the aqueous medium. Groundwater interacting closely with arsenic-rich sediments is a notable source of natural arsenic pollution. Table 7.2 shows the environmental forms of arsenic and Figure 7.3 illustrates the environmental chemistry of arsenic.

Table 7.1. Global arsenic contamination in ground water (Nordstrom, 2002)

Country/Region	Potential Exposed Population	Concentration (µg/L)	Environmental Conditions
Bangladesh	30,000,000	<1 to 2500	Natural; alluvial/deltaic sediments with high phosphate, organics
West Bengal, India	6,000,000	<10 to 3200	Similar to Bangladesh
Vietnam	>1,000,000	1 to 3050	Natural; alluvial sediments
Thailand	15,000	1 to >5000	Anthropogenic, mining, dredged alluvium
Taiwan†	100,000 to 200,000	10 to 1820	Natural; coastal zones, black shales
Inner Mongolia	100,000 to 600,000	<1 to 2400	Natural; alluvial and lake sediments; high alkalinity
Xinjiang, Shanxi	>500	40 to 750	Natural; alluvial sediments
Argentina	2,000,000	>1 to 9900	Natural; loess and volcanic rocks, thermal springs; high alkalinity
Chile	400,000	100 to 1000	Natural and anthropogenic volcanogenic sediments; closed basin; lakes, thermal springs, mining
Bolivia	50,000	-	Natural; similar to Chile and parts of Argentina
Brazil	-	0.4 to 350	Gold mining
Mexico	400,000	8 to 620	Natural and anthropogenic; volcanic sediments, mining
Germany	-	<10 to 150	Natural; mineralized sandstone
Hungary, Romania	400,000	<2 to 176	Natural; alluvial sediments, organics
Spain	>50,000	<1 to 100	Natural; alluvial sediments
Greece	150,000	-	Natural and anthropogenic; thermal springs and mining
United Kingdom	-	<1 to 80	Mining; southwest England
Ghana	<100,000	<1 to 175	Anthropogenic & natural; gold mining
USA and Canada	-	<1 to >100000	Natural and anthropogenic; mining, pesticides, As_2O_3 stockpiles, thermal springs, alluvial, closed basin lakes, various rocks

Arsenic in Rocks and Minerals

From among the rocks, igneous rocks have relatively small concentrations of arsenic (Table 7.3). Volcanic glasses show the highest arsenic content and as shown in Figure 7.2, volcanoes are important sources of As in the environment. The other igneous rocks however, do not show any marked variations. Metamorphic rocks contain about 5 mg/kg with pelitic rocks exemplified by phyllites and shales having arsenic as much as 18 mg/kg.

Sedimentary rocks are by far the more important sources of arsenic pollution and these are known to contain arsenic in the range 5-10 mg/kg. The argillaceous sedimentary rocks have higher arsenic contents as compared to the more arenaceous types such as sandstones, the values being 13 mg/kg for the former and 4 mg/kg for the latter. As shown in Table 7.3, shales of marine origin are heavily enriched in arsenic possibly due to the presence of significant amounts of sulphur and pyrite. Organic rich-sediments also contain high levels of arsenic with some coals having arsenic up to 35,000 mg/kg.

Arsenic has great affinity for sulphur and this results in arsenic being present in significantly higher concentrations in sulphides such as pyrites. Arsenian pyrite for example is a common mineral in some ore bodies and the ability of arsenic to substitute for sulphur gives rise to high As concentrations. As shown in Table 7.4, sulphides dominate the mineralogy of arsenic. Considering the rock-forming minerals (Table 7.4) arsenic, in view of its ability to substitute for Si^{4+}, Al^{3+}, Fe^{3+} and Ti^{4+}, tends to be present in many mineral structures.

It is known that authigenic pyrite is present in significant abundance in sediments of rivers, lakes and oceans as well as in aquifers. These minerals had been formed under reducing environmental conditions and they form an important link in the geochemical cycle of arsenic.

Arsenic in Soils

As shown in Table 7.3, the concentrations of arsenic in soils range from 5 to 10 mg/kg. Due to the presence of sulphide minerals, organic-rich soils tend to have higher concentrations of arsenic. Peats and bogs, for example have an average As concentration of 13 mg/kg. In Vietnam, some acid sulphate soils contain as much as 41 mg/kg. These acid sulphate soils are formed by the oxidation of pyrite in sulphide-rich terrains. It is also worthy of note that anthropogenic influences particularly from the fertilizers, also provide an input into the arsenic budget in soils.

Table 7.2. Environmental forms of arsenic (Braman, 1975)

Compound	Source
Water	
As(III), arsenite ion and	Sea water
As(V), arsenate ion	Fresh water ponds, rivers, lakes
$CH_3AsO(OH)_2$	Sea water, fresh water ponds, rivers, lakes
	Sea water, fresh water
$(CH_3)_3As$ (or other oxides)	Fresh water
Air	
As(III) and As(V)	Particulate
CH_3AsH_2	Over As-treated soil
$(CH_3)_2AsH$	Over treated soil
$(CH_3)_3As$	Over treated soil
Biological samples	
Types	Forms
Sea weed and epiphytes	As(III), As(V), $CH_3AsO(OH)_2$, $(CH_3)_2AsO(OH)$, $(CH_3)_3As$
Urine	As(III), As(V), $CH_3AsO(OH)_2$, $(CH_3)_2AsO(OH)$
Methanobacterium cultures	$(CH_3)_2AsH$
Aerobic cultures (Fungi and mixed)	$(CH_3)_3As$, $CH_3AsO(OH)_2$, $(CH_3)_2AsO(OH)$,

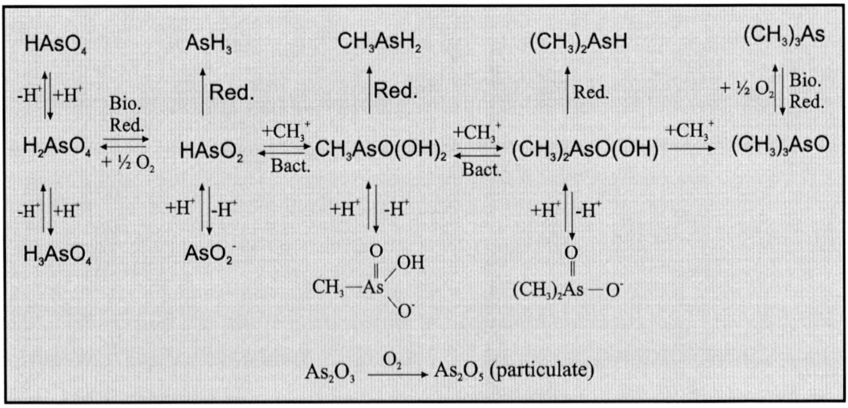

Fig. 7.3. Environmental chemistry of arsenic (Braman, 1975)

Table 7.3. Typical arsenic concentrations in rocks, sediments, soils and other surficial deposits (Smedley and Kinniburgh, 2002)

Rock / Sediment Type	Avg. Conc. (Range) (mg/Kg)
Igneous rocks	
Ultrabasic rocks (peridotite, dunite, kimberlite etc.)	1.5 (0.03-15.8)
Basic rocks (basalt)	2.3 (0.18-113)
Basic rocks (gabbro, dolerite)	1.5 (0.06-28)
Intermediate (andesite, trachyte, latite)	2.7 (0.5-5.8)
Intermediate (diorite, granodiorite, syenite)	1.0 (0.09-13.4)
Acidic rocks (rhyolite)	4.3 (3.2-5.4)
Acidic rocks (granite, aplite)	1.3 (0.2-15)
Acidic rocks (pitchstone)	1.7 (0.5-3.3)
Volcanic glasses	5.9 (2.2-12.2)
Metamorphic rocks	
Quartzite	5.5 (2.2-7.6)
Hornfels	5.5 (0.7-11)
Phyllite / slate	18 (0.5-143)
Schist / gneiss	1.1 (<0.1-18.5)
Amphibolite and greenstone	6.3 (0.4-45)
Sedimentary rocks	
Marine shale / mudstone	3-490
Shale (Mid-Atlantic Ridge)	174 (48-361)
Non-marine shale/mudstone	3.0-12
Sandstone	4.1 (0.6-120)
Limestone / dolomite	2.6 (0.1-20.1)
Phosphorite	21 (0.4-188)
Iron formations and Fe-rich sediment	1-2900
Evaporites (gypsum / anhydrite)	3.5 (0.1-10)
Coals	0.3-35,000
Bituminous shale (Kupferschiefer, Germany)	100-900
Unconsolidated sediments	
Various	3 (0.6-50)
Alluvial sand (Bangladesh)	2.9 (1.0-6.2)
Alluvial mud / clay (Bangladesh)	6.5 (2.7-14.7)
River bed sediments (Bangladesh)	1.2-5.9
Lake sediments, Lake Superior	2.0 (0.5-8.0)
Lake sediments, British Colombia	5.5 (0.9-44)
Glacial till, British Colombia	9.2 (1.9-170)
World average river sediments	5
Stream and lake silt (Canada)	6 (<1-72)
Loess silts, Argentina	3-18
Continental margin sediments	2.3-8.2

	Soils
Various	7.2 (0.1-55)
Peaty and bog soils	13 (2-36)
Acid sulphate soils (Vietnam)	6-41
Acid sulphate soils (Canada)	1.5-45
Soils near sulphide deposits	126 (2-8000)
Contaminated surficial deposits	
Mining-contaminated lake sediment	342 (80-1104)
Mining-contaminated reservoir sediment, Montana	100-800
Mine tailings, British Colombia	903 (396-2000)
Soils and tailings-contaminated soil, UK	120-52,600
Tailings-contaminated soil, Montana	up to 1100
Industrially polluted inter-tidal sediments, USA	0.38-1260
Soils below chemicals factory, USA	1.3-4770
Sewage sludge	9.8 (2.4-39.6)

Arsenic in natural waters

Smedley and Kinniburgh (2002), Matschullat (2000) and Furguson and Gavis (1972) have written extensive reviews on the geochemistry of arsenic in water and the reader is referred to these articles for further detailed information. In the natural aquatic systems, As(V) and As(III) are the dominant arsenic species. The As(V) is an oxidant, which exists as $H_2AsO_4^-$, $HAsO_4^{-2}$ and AsO_4^{-3} and under mildly reducing conditions (e.g. Eh <+100 mV), the As(III) state is stable and occurs as H_3AsO_3, $H_2AsO_3^-$, and $HAsO_3^-$ (Appelo and Postma, 1994).

The solubility of arsenic is particularly known for its strong dependence on pH and is mobilized at pH values normally found in groundwaters (pH 6.5-8.5) and under both oxidizing and reducing conditions. Figure 7.4 illustrates the Eh-pH diagram for aqueous arsenic species in the system at 25 °C and at 1 bar total pressure and Figure 7.5 shows the arsenic speciation as a function of pH. Since arsenic is a strongly redox sensitive element, its mobility in natural systems is strongly influenced by the oxidation state in which it occurs. Some of the salient features of the arsenic aqueous chemistry as discussed by Smedley and Kinniburgh (2002) are as follows:

Medical Geology of Arsenic

Table 7.4. Typical arsenic concentrations in common rock-forming minerals (Smedley and Kinniburgh, 2002)

Mineral	As conc. range (mg/kg)
Sulphide minerals	
Pyrite	100-77000
Pyrrhotite	5-100
Marcasite	20-600
Galena	5-10000
Sphalerite	5-17000
Chalcopyrite	10-5000
Oxide minerals	
Hematite	up to 160
Fe oxide (undifferentiated)	up to 2000
Fe(III) oxyhydroxide	up to 76000
Magnetite	2.7-41
Ilmenite	<1
Silicate minerals	
Quartz	0.4-1.3
Feldspar	<0.1-2.1
Biotite	1.4
Amphibole	1.1-2.3
Olivine	0.08-0.17
Pyroxene	0.05-0.8
Carbonate minerals	
Calcite	1-8
Dolomite	<3
Siderite	<3
Sulphate minerals	
Gypsum/anhydrite	<1-6
Barite	<1-12
Jarosite	34-1000
Other minerals	
Apatite	<1-1000
Halite	<3
Fluorite	<2

(a) Most oxyanions including arsenate tend to become strongly sorbed as the pH increases. This is in contrast to most trace-metal cations which show low solubility due to precipitation or co-precipitation with an oxide, hydroxide, carbonate or phosphate mineral or due to strong adsorption to metal oxides, clay or organic matter. Arsenic is thus an important trace contaminant in groundwaters.

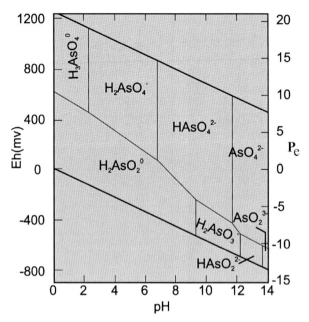

Fig. 7.4. Eh-pH diagram for aqueous As species in the system As-O$_2$-H$_2$O at 25°C and 1 bar total pressure (Smedley and Kinniburgh, 2002)

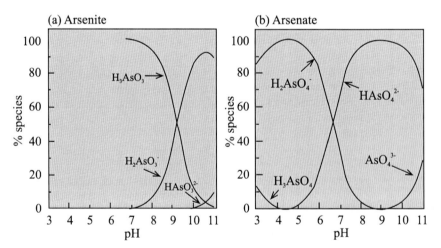

Fig. 7.5. (a) Arsenite and (b) arsenate speciation as a function of pH (ionic strength of about 0.01 M) (Smedley and Kinniburgh, 2002)

(b)   Arsenic shows marked mobility under reduced conditions. It can be found at concentrations in the mg/L range when all other oxyanion-forming elements are present in the µg/L range.

Medical Geology of Arsenic

(c) When strongly reducing acidic conditions prevail, precipitation of sulphides such as orpiment (As_2S_3) and realgar (AsS) occur resulting in low-arsenic waters.

(d). The concentrations of arsenic in groundwaters vary markedly depending on the source and amount of arsenic as well as the geochemical environment. When the water-rock interactions exert a strong influence, As concentrations reach high levels, in the aqueous phase due to increased mobilization and accumulation.

(e). In atmospheric precipitation, arsenic is mainly in the form of $As(III)_2O_3$ dust particles. As shown by Nriagu and Pacyna (1988), anthropogenic sources of atmospheric arsenic amounts to 70% of the global atmospheric As flux.

(f). The arsenic concentrations of river water vary depending on the recharge composition, base flow contribution and bedrock lithology. In areas where there are high geothermal inputs, high arsenic concentrations can be expected. In some arid areas where surface water is dominated by river base flow, high arsenic levels may be observed and these often correlate with salinity.

(g). The oxidation and adsorption of As species onto river sediments and the dilution by surface recharge and run off tends to lower the arsenic concentration in the river water even if the bed rock may show high amounts of As.

(h). River waters can also be affected by high arsenic inputs from anthropogenic sources, bearing in mind that As is a well known industrial contaminant.

(i). In lake water arsenic concentrations tend to show evidence of stratification as a result of varying redox conditions. The As(III) to As(V) ratio may increase with depth. Higher arsenic concentrations are also observed at deeper lake waters when O_2 is depleted due to influence of microbial activity.

(j). In groundwater, there is a large range of arsenic concentrations reported in the literature. Most high-As groundwater provinces are the result of natural occurrence of As.

A review of the geochemical processes that control arsenic mobility indicates that the main processes of arsenic mobility in aquifers is (a) adsorption and desorption reactions (b) solid-phase precipitation and dissolution reactions. Arsenic adsorption and desorption reactions are influenced by pH changes, redox reactions, presence of competing anions and solid-phase structural changes at the atomic level.

Arsenic Adsorption and Desorption: Role of Fe, and Mn Oxides and Clays

Arsenate and arsenite adsorb onto surfaces of materials in the aqueous medium, and most notably onto iron oxides, aluminium oxides and clay minerals. As observed by Dzombak and Morel (1990) adsorption and desorption reactions between arsenate and iron oxide surfaces are of great importance as controlling reactions. Iron oxides are ubiquitous in the aqueous environment as coatings on solids (Figure 7.6). Arsenate adsorbs very strongly to these iron oxide surfaces in acidic and near-neutral pH water and desorbs as pH increases.

Iron oxide surfaces are also known to adsorb arsenite, and both arsenate and arsenite adsorb onto aluminium oxides and clay mineral surfaces. Under the typical environmental pH conditions these adsorption-desorption reactions however, appear to be less active than those between arsenate and iron-oxides surfaces.

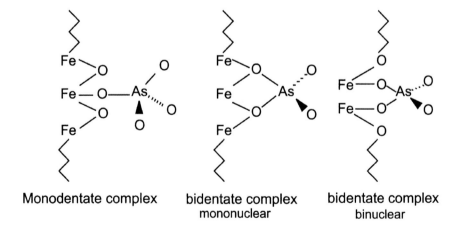

Fig. 7.6. Schematic illustration of the proposed adsorption mechanisms of arsenate onto goethite (O'Reilly et al., 2001)

Recent work (Goldberg and Johnston, 2001), using spectroscopic, sorption and electrophoretic mobility measurements show that arsenate forms inner sphere surface complexes (Figure 7.7) on both amorphous Al and Fe oxides while arsenite forms both inner and outer sphere surface complexes on amorphous Fe oxides and outer sphere surface complexes on amorphous Al oxide. Arsenic adsorption on amorphous Al and Fe-oxides and the clay minerals, kaolinite, montmorillonite and illite was investigated by Goldberg (2002) as a function of solution pH and As redox state, i.e., As(III) and As(V) (Figure 7.8). The results indicated that arsenate adsorption on oxides and clays was maximal at low pH and decreased with increasing pH above pH 9 for Al oxides, pH 7 for Fe oxide and pH 5 for clays. Arsenite adsorption exhibited parabolic behaviour with an adsorption maximum around pH 8.5 for all materials. There was no competitive effect of the presence of equimolar arsenite on arsenate adsorption.

Microorganisms and their impact on arsenic speciation and mobility

It is well known that microorganisms are ubiquitous in the geochemical environment and that they influence the biogeochemical cycles of many elements. Arsenic is such an element that inter-converts species which show markedly contrasting behaviour. The role of microorganisms is now of special importance in view of the extremely large scale of arsenic poisoning in West Bengal and Bangladesh. Many theories have been proposed to explain the subsurface mobilization of arsenic. Among these are:

(a) Oxidation of arsenic containing pyrites.
(b) Release of As(V) from reduction of iron oxides by autochthonous organic matter such as peat.
(c) Reduction of iron oxides by allochthonous organic matter (from dissolved organics in recharging waters)
(d) Exchange of adsorbed As(V) with fertilizer phosphates.

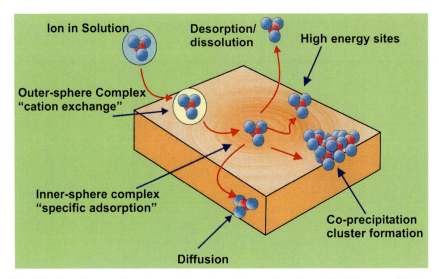

Fig. 7.7. Sorption mechanism at the mineral/water interface (Goldberg and Johnston, 2001)

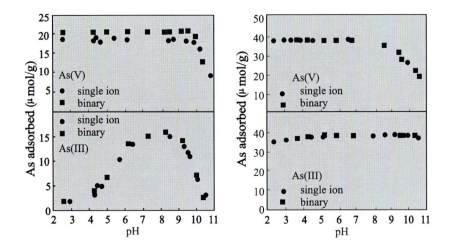

Fig. 7.8. Arsenic adsorption on kaolinite and montmorillonite as a function of pH and As redox state. Single ion layer systems: AsΓ -20µM, Binary system As(III)Γ –As(V)Γ- 20 µM. Suspension density 40 g/L (Goldberg and Johnston, 2001)

Oremland and Stolz (2003) have suggested that the above processes are not mutually exclusive, but that over time, microorganisms probably play an essential role in both the direct reduction and oxidation of the arsenic species, as well as the iron minerals contained in the aquifers. Microorganisms catalyze chemical reactions in order to obtain energy for metabolic growth. They obtain this much needed energy from the environment and as a result of the chemical reactions catalyzed by the microorganisms, extremely important biochemical transformation may occur. The greatest energy-generating reactions of the biosphere are redox reactions and these cause marked alterations in the behaviour of their substrates (Stumm and Morgan, 1996). The great need for microbes to catalyze the energy-generating reactions causes rapid chemical transformation in contrast to the somewhat sluggish rates of abiotic reactions. The microorganisms catalyze the redox reactions by means of aerobic and anaerobic respiration and these processes influence the arsenic biogeochemistry.

The discovery that in spite of its toxicity, As(V) is actually used as a respiratory oxidant had opened up further avenues for microbial biogeochemical research. Two types of related bacteria, namely *Sulfurospirillium arsenophilum* and *Sulfurospirillium barresii* were the first microbes reported that carried out this process (Ahman et al., 1994; Oremland and Stolz, 2003). Such microbes are termed dissimilatory arsenate-reducing prokaryotes (DARPs) and have been isolated from fresh water sediments, estuaries, soda lakes, hot springs and even gold mines (Oremland et al., 2002). Figure 7.9 illustrates the diversity of the arsenic-metabolizing prokaryotics. It is of interest to note that some of these microbes have been isolated from the aquifer materials of Bangladesh and gastrointestinal tract of animals. There are also the extremophiles adapted to high temperature, pH and/or salinity (Oremland and Stolz, 2003). These authors formulated a conceptual model for the geochemical scenario of Bangladesh arsenic-rich aquifer systems (Figure 7.10).

The sequence of the mechanism proposed is as follows:

i) Oxidation of the original As(III) containing minerals. (eg. Arseno pyrite) during transport and sedimentation by pioneering chemo lithoautotrophic arsenite oxidizers (CAOs) and heterotrophic arse nate oxidizers (HAOs) taking place over recent geological time periods.

ii) Accumulation of As(V) onto surfaces of oxidized minerals such as ferrihydrite.

iii) Later anthropogenic activities such as irrigated agriculture, digging of wells, lowering of groundwater table by water extraction etc.

would provide oxidants (e.g. oxygen, nitrates) which would stimulate As(III) oxidation.

iv) This would cause a build up of microbial mass and hence organic matter thereby creating anoxic conditions.

v) These and other organic matter such as peat would promote the dissimilatory reduction of absorbed As (V) by DARPs and the subsequent dissolution of absorbent minerals such as ferrihydrite.

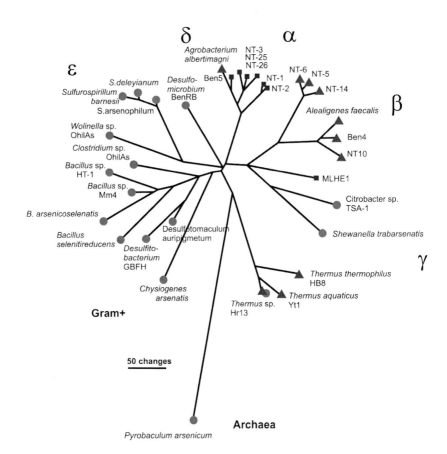

Fig. 7.9. Phylogenetic diversity of representative arsenic-metabolizing prokaryotics. Dissimilatory arsenate-respiring prokaryotes (DARPs) are indicated by yellow circles, heterotrophic arsenite oxidizers (HOAs) are indicated by green triangles, and chemoautotrophic arsenite oxidizers (CAOs) are indicated by red squares. In some cases (e.g. *Thermus sp.* strain HR13), the microbe has been found able to both respire As(V) and oxidize As(III) (Oremland and Stolz, 2003; reprinted with kind permission from AAAS)

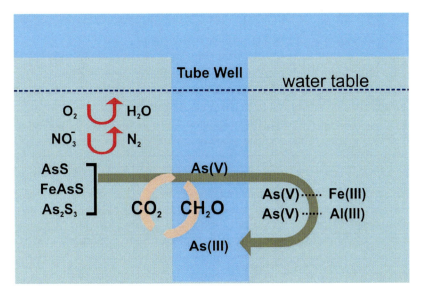

Fig. 7.10. A conceptual model of how arsenic-metabolizing prokaryotes may contribute to the mobilization of arsenic from the solid phase into the aqueous phase in a subsurface drinking water aquifer. *Arsenic is originally present primarily in the form of chemically reduced minerals, like realgar (AsS), orpiment (As_2S_3), and arsenopyrite (FeAsS). These minerals are attacked by CAOs, which results in the oxidation of As(III), as well as iron and sulfide, with the concurrent fixation of CO2 into organic matter. Construction of wells by human activity accelerates this process by providing the necessary oxidants like molecular oxygen or, in the case of agricultural regions, nitrate. The As(V) can subsequently be adsorbed onto oxidized mineral surfaces like ferrihydrite or alumina. The influx of substrate organic materials derived either from buried peat deposits, recharge of surface waters, or the microbial mats themselves promotes microbial respiration and the onset of anoxia. DARPs then respire adsorbed As(V), resulting in the release of As(III) into the aqueous phase* (Oremland and Stolz, 2003; reprinted with permission from AAAS)

The research studies of Saikat et al. (2001) also show evidence of microbial activity which mediate in arsenic transformations in the Bangladesh aquifer sediments.

The bacterial dissimilatory reduction of arsenate and sulphate has been studied in detail by Oremland et al. (2000) in Mono Lake California. The stratified (meromictic) water column of alkaline and hypersaline Mono Lake contains high concentrations of dissolved inorganic arsenic (~200 μmol/L). Arsenic speciation changed from As(V) to As(III) with the transition from oxic surface waters (mixolimnion) to anoxic bottom waters (monolimnion). A radio assay devised to measure the reduction of ^{73}As(V) to ^{73}As(III) and tested using cell suspensions of the As(V)-respiring *Bacillus selenitireducens* showed complete reduction of ^{73}As(V).

Anoxic environments host a variety of microorganisms and these generate energy by coupling the oxidation of H_2 or organic carbon to the reduction of inorganic As(V) to As(III). Because arsenate reduction is an energy-generating process for the microorganisms involved, and because arsenite is both toxic and more mobile than arsenate, this process is expected to greatly influence the geochemistry of arsenic in anoxic systems, particularly with respect to arsenic mobilization (Ahman et al., 1997; Cummings et al., 1999).

As discussed earlier, sorption onto oxides of Fe and Mn and precipitation on sulphide solids in anoxic environments are the two main processes that influence arsenic mobility in aqueous, soil and sedimentary environments. Microbial impacts on Fe and Mn oxides therefore influence arsenic geochemistry. Since the reduced forms of Fe and Mn are highly soluble, these oxides dissolve readily upon microbial reduction, simultaneously releasing sorbed substances such as arsenic (Ahman et al., 1997). Further, microbial sulphate reduction can yield sufficient sulphide to precipitate arsenic in solids such as amorphous arsenic sulphide, realgar or orpiment. While the geochemical significance of the reductive processes for arsenic cycling is well-established, the significance of the oxidative pathways is much less understood.

Apart from energy generation, microorganisms need to protect themselves from toxic substances. The mechanism of arsenic toxicity is for arsenate to enter the microbial cell through phosphate uptake proteins facilitated by the structural of similarity of arsenate to phosphate. The phosphate in the Adenosine Triphosphate (ATP) is displaced by the arsenate thereby reducing its energy. It should however be noted that there are many cells which are highly phosphate-specific and which exclude arsenate (Torriani, 1990). Arsenite on the other hand enters the cytoplasm possibly by diffusion across the membrane and then cross-links sulphydryl groups on enzyme. These cause the inactivity of the enzyme.

The mechanisms adopted by the microorganisms to remove the effects of arsenic poisoning are of special interest. The microbial reduction of arsenate to arsenite is by means of the As system an enzymatic process. Other mechanisms involve detoxification by reduction to arsine, As(III) in inorganic and methylated forms (Cullen and Reimer, 1989). Extensive studies are now being done on these microbial arsenic detoxification mechanisms in order to utilize them in cleaning up polluted environments.

MEDICAL GEOLOGY OF ARSENIC- THE WEST BENGAL, BANGLADESH EXAMPLE

The arsenic poisoning in Bangladesh caused by high As concentration in aquifers is the largest mass poisoning in history. With up to 125 million of Bangladesh's population at risk, the case of arsenic poisoning in Bangladesh, and West Bengal is a classic example of the importance of the understanding of medical geology.

Bangladesh Basin-Geography and Geology

The geological, geographical, sedimentological and hydrological feature of the Bangladesh Basin, particularly with reference to the arsenic problem has been discussed exhaustively in the British Geological Survey Report of 2001 and the reader is referred to this valuable source of information for details. Only a brief outline is presented here.

The Bengal Basin comprising of a 15 km thick sequence of Cretaceous to Recent sediments and occupying about 100,000 km^2 of lowland flood plain and delta is one of the largest in the world. The combined deltas of the Ganges, Brahmaputra and Meghna (GMB) river systems, all of which lie within Bangladesh produce the greatest sediment load of any river system in the world (Figure 7.11). The sediment loads are known to vary by two orders of magnitude seasonally and these are mostly derived from the glacial and periglacial activity of the high Himalayas. Among the rocks that had undergone erosion are the ultrabasics of the northern high Himalayas and the granitic and high-grade metamorphic rocks from the central and southern parts. The extensive delta system has developed as a result of a series of glacio-eustatic sea level cycles and long-term tectonic activity. The very large number of aquifer systems underlying Bangladesh is a result of the deposition of Pleistocene to Recent fluvial and estuarine sediments. As shown in Figure 7.12, the surficial materials comprise mainly of alluvium, clay, muds, gravels, sand and peat.

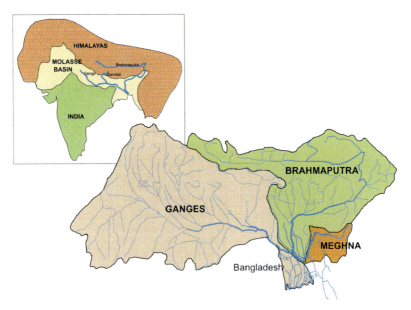

Fig. 7.11. Ganga-Brahmaputra-Meghna (GMB) river systems

Sediment characteristics

The British Geological Survey (2001) using limited borehole data constructed a hydrogeological cross section from north to south across Bangladesh and series of cycles of sedimentation has been observed. In the zone of subsidence at the northern end is a wedge of coarse sediments deposited as fanglomerates. In the zone of uplift at Rangpur saddle the sediments thin out. The coarse-grained sediments thin and pinch out south of Hinge Zone and pass laterally into sandy deltaics within the subsiding Faridpur Trough. Towards the south in the coastal zone, a series of aquifers is seen with alternations of sandstones and silts. The aquifers located away from the saline intrusions have been united to form a single body of fresh water.

Hydrochemical and mineralogical studies had revealed that the sediments containing groundwaters with the highest concentration of arsenic are the shallow fine-grained highstand deposits which had radiocarbon dates generally less than 10 ka. These are found to be localized in the tidal zones of the present Ganges, old Ganges, lower Meghna and lower Brahmaputra delta areas.

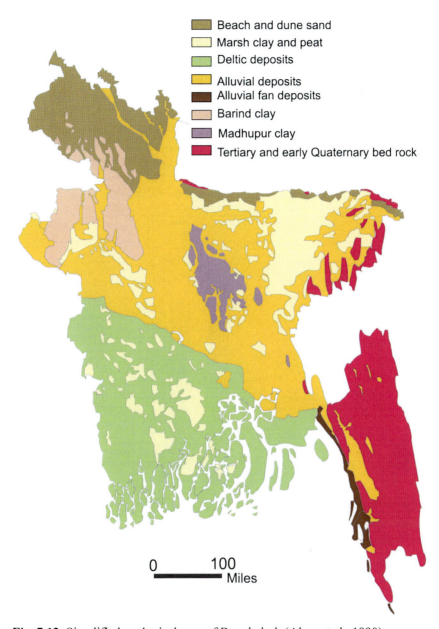

Fig. 7.12. Simplified geological map of Bangladesh (Alam et al., 1990)

Mineralogy and geochemistry of sediments

The mineralogy of the aquifer sediments have been studied by a large number of investigators, in different parts of the Bengal basin. Pal et al. (2002) studied the mineralogical constituents in two bore holes in the Chakdah area, West Bengal, and observed that clay and peat layers are richest in arsenic while the silty sand layers contain lesser amounts of arsenic. Figure 7.13 illustrates the distribution of Fe and As in two sediment cores beneath Deulhi Village (D3) and Smata Village (D4) near Jessore, south western Bangladesh. Peat has been observed (Ravenscroft et al., 2001).

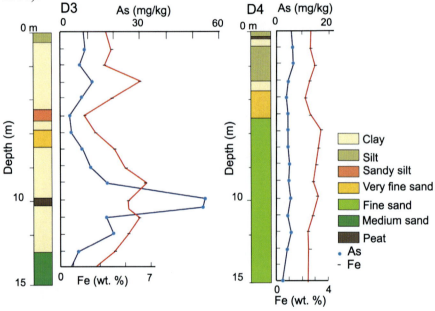

Fig. 7.13. Distribution of Fe and As in two sediment cores beneath Deulhi Village (D3) and Samta Village (D4) near Jessore, SW Bangladesh showing the occurrence of peat (Ravenscroft et al., 2001)

Pal et al. (2002) studied arsenious zone and 'safe water zone' of the West Bengal sections, and noted that the sandy aquifer material underlying the clay, peat and silty layers in both arsenic-rich and arsenic-poor zones had silt to sand-sized grains (96-99%) and interstitial clay (1-4%). The coarser components consisted of (a) a non-magnetic fraction such as stain-free quartz, feldspar, carbonate and lithic fragments and (b) minerals with variable magnetic intensity constituting 12-36% and which included (i) strongly magnetic phases such as iron oxide/hydroxide with residual magnetite and ilmenite (ii) feebly magnetic phases such as illite, biotite, iron oxyhydroxide-coated sand grains, siderite concretion, muscovite, garnet, hornblende, tourmaline, rutile, kyanite, sillimanite, epidote, apatite, staurolite, zircon. The finer interstitial clay fraction contained e.g. quartz, feldspar, kaolinite, illite, montmorillonite and chlorite.

An important feature worthy of special note, is the total absence of arsenic-bearing clastic phases such as arsenopyrite and arsenious pyrite. The British Geological Survey (2001) also noted that arsenopyrite was never seen and that in general, sediments are sulphur-poor. They confirmed the general observation that the highest concentrations of most elements, including arsenic, are found in the fine-grained, especially clay, sediments. The hypothesis that iron oxides are a primary source of As in Bangladesh groundwaters was supported by the BGS data. Areas which had high concentrations of iron oxides with their adsorbed and co-precipitated As had the most arsenic-rich ground waters.

Organic matter

The presence of organic matter, notably peat, in some of the aquifer sediments of the aquifer sediments of the Bengal Basin has influenced some workers to hypothesize on the role of organic matter in the arsenic geochemistry. Ahmed et al. (1998); Brammer (1996), Safiullah (1998), DPHE (1999), Ishiga et al. (2000) noted the presence of peat beds in deferent parts of the Bengal Basin. Further locations of peat are given in Ravenscroft et al. (2001), who report that biogenic methane is common in groundwater over large areas, in places in amounts sufficient to provide domestic fuel. This was taken as a clear indicator of a substantial amount of organic matter undergoing microbial degradation. Human organic wastes from latrines, some of which are located very close to wells are also considered as contributors to pollution. Ravenscroft et al. (2001) considered peat degradation as a major redox driver of arsenic pollution (Figure 7.14).

The scale of the problem

Termed as the biggest mass poisoning in history, the arsenic problem in the Bengal Basin, in view of the sheer numbers of people involved (estimated to be over 130 million) is by far the biggest arsenic groundwater problem in the world. Up to half of Bangladesh's tube wells (Figure 7.15), about 10 million in number, are thought to be contaminated and the expected deaths resulting from arsenic poisoning is expected to run into thousands. It is estimated that 95% of the population of Bangladesh use groundwater as drinking water. It is the shallow aquifer system (10-70 m below ground level) that is known to be the main cause of the arsenic problem. The deep aquifer system (150-200 m) is known to have relatively less arsenic.

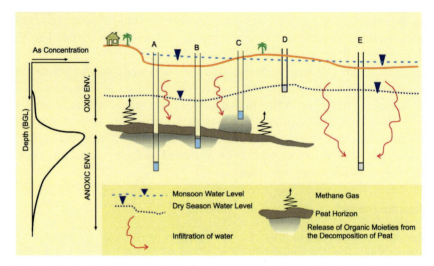

Fig. 7.14. Model of how arsenic pollution occurs in Bengal Basin and in any sedimentary sequence hosting buried swampland and marsh. In shallow Bangladesh sequences, hydraulic gradients cause downward movement of water during the wet season. In other sequences hydraulic gradients may cause upward flow of water through peat (Ravenscroft et al., 2001)

Well A- Low concentrations of organic moieties from distant peat cause some FeOOH reduction, the release of small amounts of arsenic and so low arsenic concentrations.

Well B- High amount of organic moieties from near by peat cause much FeOOH reduction, the release of large amount of arsenic and high arsenic concentrations.

Well C- Arsenic pollution above a peat layer caused by migration of arsenic in response to strong pumping and also migration of organic moieties upwards to cause local FeOOH reduction and additional arsenic release.

Well D- Uncontaminated, oxic, hand dug well. Seasonally dry and is safe from arsenic pollution.

Well E- A well that is currently uncontaminated. The likelihood of contamination depends on the distance organic moieties travelling laterally before being consumed by redox reactions and on the rate of movement of dissolved arsenic.

Fig. 7.15. There are millions of tube wells in Bangladesh, providing 95% of the drinking water in this country

The geochemical mechanisms of arsenic mobility in the Bengal basin

Distribution of Arsenic in the Aquifer System

The extensive work carried out by the British Geological Survey in 41 of the 64 districts in 1998 showed that from among 2022 samples analyzed

i) 51% of the samples were above 10 µg/L (WHO guideline value)
ii) 35% were above 50 µg/L
iii) 25% were above 100 µg/L
iv) 8.4% were above 300 µg/L
v) 0.1% were above 1000 µg/L

Only about 20% were considered to be essentially 'arsenic-free'.

The survey also showed that the groundwaters are characteristic of reduced waters as indicated by high dissolved Fe, Mn and low sulphate. Unusually high phosphate concentrations were also observed (median 0.6 mg/L). Figure 7.16 illustrates the map showing the distribution of arsenic in groundwater in Bangladesh. The occurrences of arsenic correlates strongly

182   Introduction to Medical Geology

with surface geology and geomorphology, the most affected aquifers being those alluvial deposits lying beneath the Recent flood plains. From among the flood plains, the Brahmaputra and Tista rivers had the lowest levels while those of the Meghna flood plains southeast Bangladesh had much higher concentrations.

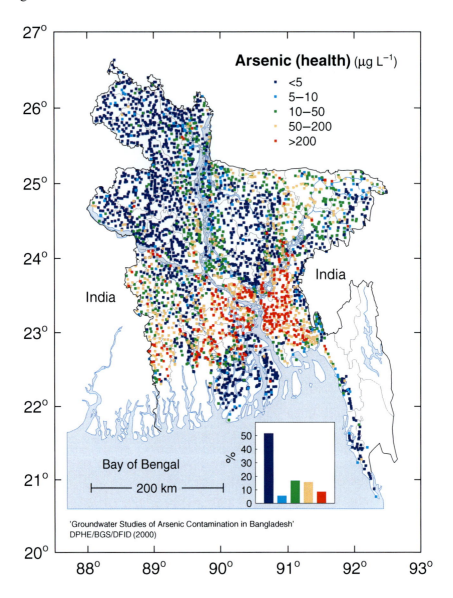

Fig. 7.16. Distribution of arsenic in groundwater in Bangladesh (reprinted with kind permission from the British Geological Survey, 2001)

Geochemical Mechanism of Arsenic Mobility

There are at present three main mechanisms attributed by different workers to the mechanism of enrichment of arsenic waters in the Bengal Basin. These could be summarized as:

i) **Mechanisms of As-bearing pyrite oxidation**
In this hypothesis (Mallick and Rajagopal, 1996; Das et al., 1996; Mandal et al., 1998), arsenic is released by the oxidation of arsenic-bearing pyrite in the sediments when groundwater irrigation lowers the water table to allow atmospheric oxygen into the aquifer sediments.

ii) **Competitive (anion) exchange mechanism**
The arsenite/arsenate anions sorbed to aquifer minerals are displaced into solution by competitive exchange with other oxyanions, such as phosphate silicate or bicarbonate. In the Bengal Basin (Achryya et al., 2000) the source of phosphate was considered to be the super-phosphate fertilizer that was applied in excess and which allowed the leaching of arsenic into the groundwaters.

iii). **FeOOH reduction mechanism**
Naturally occurring arsenic sorbed onto iron oxyhydroxides (FeOOH) is released when these are reduced at anoxic conditions developed during sediment burial (Bhattacharya et al. 1997; Nickson et al., 1998; 2000; McArthur et al., 2001), possibly driven by microbial actions.

The mechanism of arsenic-bearing pyrite oxidation has not found much favour among the large number of researchers investigating into the mechanism of arsenic mobilization. Arsenopyrite has not been identified among the common minerals in Bangladesh sediments (Pal et al., 2002). The lack of high sulphate concentrations in arsenic-rich groundwater is used as a key evidence against the pyrite oxidation hypothesis.

The second mechanism concerning the competitive anion exchange with phosphates and others stemmed from the increased use of phosphatic fertilizers in Bangladesh. The sorbed arsenic on FeOOH was thought to be classified as a result of phosphate leached from soils on account of an excessive use of phosphatic fertilizer. However, this idea has been rejected (Ravenscroft et al., 2001) in view of the fact that there are certain areas where the groundwater is essentially free of both arsenic and phosphorus but where irrigation and application of fertilizer is highest. Some experimental work carried out by Manning and Goldberg (1997) showed that P/As mole partition ratios for desorption of arsenic by phosphate were around 5000. Thus, no more than 2 μg/L of arsenic was expected to be desorbed by a phosphorous (as P) concentration in groundwater of 5 mg/L. Further, uranium which normally accompanies phosphorus in fertilizers was found to be very low in the Bangladesh groundwaters. Even when uranium was high, phosphorus was very low. Thus most accepted hypothesis for arsenic enrichment of Bangladesh aquifer is the iron oxide reduction hypothesis.

Arsenic is known to show slow release from recently buried sediments in rivers, lakes and ocean (Smedley and Kinniburgh, 2002). The iron oxide/hydroxide phases adsorb arsenic on their surfaces, bearing in mind that many minerals have the surface-coated iron oxide/hydroxide and which therefore becomes arsenic-rich. The reduction of FeOOH is driven by a microbial metabolism of organic matter and is accompanied by microbial reduction of arsenate to arsenite (Zobrist et al., 2000).

The desorption of the above arsenic by some process will undoubtedly enrich the groundwaters in arsenic. The oxidation of fresh organic matter during burial of sediments results in anaerobic conditions, thus aiding the release of arsenic. Arsenic release is probably due to the following processes (BGS, 2001):

(a) Reductive desorption of arsenic due to transformation of As(V) to As(III).
(b) Iron oxide reduction: reductive dissolution of iron oxides, and a change in surface structure and specific surface area of the iron oxides during diagenesis.
(c) Competition from other anions such as phosphates that may be strongly bound.

The very large volume of the sediments in the Bengal Delta causes a massive total accumulation of arsenic, even though the actual mean arsenic concentration may not be excessively high. The geochemical processes occurring in the delta causes the enrichment of arsenic-rich groundwaters.

The studies of Ravenscroft et al. (2001) and Harvey et al. (2002) indicate a distinct role played by organic matter. In their model of arsenic mobilization in the Bangladesh aquifer sediments, Ravenscroft et al. (2001) hypothesise that small organic species such as short-chain carboxylic acids and methylated amines drained from biodegradation of peat drives the FeOOH reduction and ammonium production. The strong dependence of arsenic distribution on depth was taken to suggest that these small organic molecules have not migrated far and lie close to the peat source. Their model (Figure 7.14) emphasizes the importance of the biodegradation of buried peat deposits in the extreme reduction of FeOOH and high arsenic groundwaters. Harvey et al. (2002) are also of the view that arsenic mobilization is associated with the recent inflow of carbon. High concentrations of radio-carbon methane indicated that young carbon has driven the recent biogeochemical processes.

Arsenic in Rice and Other Crops

The arsenic-enriched water in Bangladesh is used commonly to irrigate paddy (rice) fields and vegetable plots. Arsenic therefore gets into the food chain quite easily. Paddy rice (*Oryza sativa* L.) is the staple food of the millions of Bangladeshis and the nature of the uptake of arsenic by the rice plant is of special importance, bearing in mind that tens of kilograms of rice per year per person is consumed.

Meharg and Rahman (2003) carried out a survey of paddy soils and rice grains throughout Bangladesh. The survey of the paddy soils showed that where arsenic groundwater is used for irrigation, the arsenic contents were elevated. Arsenic levels in the rice grains from an area of Bangladesh with low levels of arsenic in groundwaters and in paddy soils showed that the levels were typical of other regions of the world. The rice grains grown on regions where soil arsenic was high, had markedly high arsenic contents, with three rice grain samples having as much as 1.7 µg/g As. A study by Huq et al. (2006) showed that from among the crops, Arum (*colocassia antiquorum*) and Boro rice had the highest arsenic build up (Figure 7.17).

The uptake kinetics of arsenic species in rice plants has been studied by Abedin et al. (2002). They showed that As species found in soil solution from a greenhouse experiment where rice was irrigated with arsenate contaminated water were, arsenite, arsenate, dimethylarsinic acid and monomethylarsonic acid. Competitive inhibition of uptake with phosphate showed that arsenite and arsenate were taken up by different uptake systems because arsenate uptake was strongly suppressed in the presence of phosphate, whereas arsenite transport was not affected by phosphate. Abedin et al. (2002) also showed that there was a hyperbolic uptake of monomethylarsonic acid and limited uptake of dimethyl arsinic acid, at a slow rate.

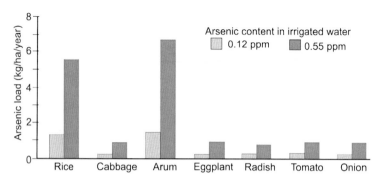

Fig. 7.17. Arsenic load from irrigation water for some crops compared to that of rice (Huq et al., 2006)

HEALTH EFFECTS OF ARSENIC

Arsenic found in the natural environment can enter the human body through food, water, soil and air. Skin contact with arsenic-rich soil and water could also be a source of irrigation. The particular arsenic species in the environment is not normally determined in routine procedures and hence the level and nature of arsenic exposure may not be known.

Arsenic is known as a protoplasmic poison due to its effect on the thiol group (-SH) of protein molecules in cells affecting normal enzymatic action, cell respiration and mitosis (Gordon and Quastel, 1948). In the biotransformation of arsenic (Scheme shown below), monomethyl arsenous acid (MMA III) is reported to be the most toxic to the cells (Aposhian et al., 2000) and a powerful inhibitor of glutathione reductase and thioredoxin reductase (TR).

$$\begin{array}{c} \text{Arsenate [As(V)]} \\ \downarrow \quad \text{Arsenate reductase} \\ \text{Arsenite [As(III)]} \\ \downarrow \quad \text{Arsenate methyl transferase} \\ \text{MMA V (Monomethyl arsonic acid)} \\ \downarrow \quad \text{Glutathione Reductase} \\ \text{MMA III (Monomethyl arsenous acid)} \\ \downarrow \quad \text{MMA methyl transferase} \\ \text{DMA V (Dimethyl arsinic acid)} \end{array}$$

Arsenic is classified as a hazardous material suspected to be a carcinogen affecting the lungs and skin. It is also a teratogen which means that it can cross the placental membrane into the metabolic system of unborn children. It is also known as a cumulative substance passing out of the body through urine, hair, finger, toe nails and skin. Most of the arsenic poisoning symptoms have been identified in Bangladesh. Among these are patients with melanosis (93.5%), leuco-melanosis (39.1%), keratosis (68.3%), hyperkertosis (37.6%), dorsum, non-petting oedema, gangrene and skin cancers (Karim, 2000). Tables 7.5 and 7.6 show the concentrations of arsenic in urine, hair and nails in some areas of the world, and the magnitude of the arsenic problem in Bangladesh, respectively. Figure 7.18 shows the impact of arsenic poisoning on the health of people living in affected areas. Saha (2003) in a review of arsenicosis in West Bengal discussed the severity of the arsenic problem in West Bengal. Arsenicosis is classified into 4 stages (i) preclinical (ii) clinical (iii) complication and (iv) malignancy (Saha et al., 1999) (Tables 7.7 and 7.8).

Table 7.5. Concentration of arsenic in urine, hair and nails of the affected people in different arsenic contaminated water ingestion episodes (Karim, 2000)

Location	Concentration in urine (mg/L)	Concentration in hair (mg/kg)	Concentration in nails (mg/kg)
Fairbank (USA)	0.18	1.0	4.0
Millard Country (USA)	0.025-0.66	010-4.7	-
Antofagasta Chile	0.025-0.77	4.00-83.4	-
Lassen Country (USA)	0.04-0.26	0.01-2.00	-
Taiwan	0.03-2.0	-	-
West Bengal, India	0.03-2.0	1.81-31.05	1.47-52.03
Bangladesh	0.05-9.42	1.1-19.84	1.3-33.98

Table 7.6. Arsenic in different body tissues collected from Bera, Ishurdi and Kushita areas-Bangladesh (the urine arsenic content were estimated on the assumption that a total discharge of urine at one day is 1.5L) (Karim, 2000)

Sample	Total Samples	Normal Category No.	Normal Category %	Safety range	Higher than normal No.	Higher than normal %	Arsenic conc.
Finger-nails	74	4	5	0.43-1.08 mg/kg	70	95	1.3-33.98 mg/kg
Hair	74	3	4	0.08-0.25 mg/kg	71	96	1.1-19.84 mg/kg
Skin	65	0	0	na	65	100	0.28-23.5 mg/L
Urine	63	4	6	0.005-0.04 mg/dl	59	94	0.075-14.13 mg/dl
Water	41	14	34	0.01-0.05 mg/L	27	66	0.01-9.0 mg/L

Table 7.7. Stage-wise gradation of arsenicosis (Saha et al., 1999)

Stages	Grades	Inference
I. Preclinical	0	Preclinical
	0-a	Labile-Blood phase
	0-b	Stable-Tissue phase
II. Clinical	1	Melanosis
	1-a	Diffuse melanosis in palm
	1-b	Spotted melanosis in trunk
	1-c	Generalized melanosis
	2	Spotted keratosis in palms and soles
	2-a	Mild (1- 6 nodules)
	2-b	severe (>6 nodules)
	2-c	large nodules
	3	Diffuse keratosis in palms and soles
	3-a	Partial- in palms or soles
	3-b	Partial- in palms and soles
	3-c	Complete
	4	Dorsal keratosis
	4-a	In hands or legs
	4-b	In hands and legs
	4-c	Generalized
III. Complications	5	Hepatic disorder
	5-a	Palpable liver
	5-b	Jaundice
	5-c	Ascitis
IV. Malignancy	6	Malignancy
	6-a	Single lesion
	6-b	Two lesions
	6-c	More than two lesions

Table 7.8. Increasing incidences of arsenicosis in West Bengal (1983-1998) (* estimated figure) (Saha et al., 1999)

Year	Affected districts	Affected blocks	Affected villages	Arsenical Dermatitis patients
1983	4	5	5	127
1984	5	12	15	241
1985	6	17	24	485
1986	6	30	40	1068
1987	6	40	61	1214
1988	6	42	78	2026
1989	6	43	79	2185
1990	6	44	123	24000*
1993	6	47	415	83000*
1994	6	47	428	85600*
1995	6	54	544	108800*
1996	7	60	638	200000*
1997	9	74	966	>200000*
1998	9	76	1206	>225000*

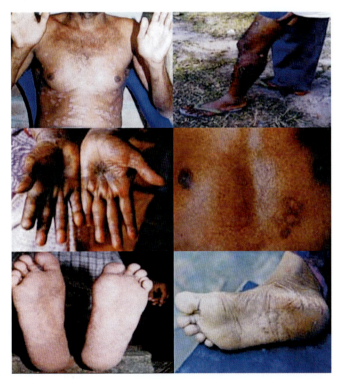

Fig. 7.18. Impact of arsenic poisoning on the health of people living in affected areas of West Bengal (photos sent by Prof. Surendra Kumar)

CHAPTER 8

WATER HARDNESS IN RELATION TO CARDIOVASCULAR DISEASES AND URINARY STONES

One of the most intriguing, yet, not very well defined geochemistry-health correlations is the incidence of cardiovascular diseases (CVD) in connection with water hardness of a particular area (Crawford et al., 1977; Comstock, 1979; Bernardi et al., 1995). One of the earliest studies on the relationship between water hardness and the incidence of vascular diseases was by a Japanese chemist Kobayashi (1957). He showed on epidemiological grounds higher mortality rates from cardiovascular diseases (strokes) in the areas of Japanese rivers with softer water compared to areas with hard water used for drinking purposes. Kožíšek (2003) has summarized the beginnings of research that led to the health significance of water hardness. He mentions that among the best known studies is that by Schroeder (1960) who showed a correlation between mortality from CVD in males (ages 45-64) and water hardness in 163 largest cities of the USA and who summarized his results under the caption **"soft water, hard arteries"**.

Within the first two decades of research into water hardness in association with cardiovascular diseases, more than 100 papers had been published (Hewitt and Neri, 1980). In several countries and areas, a negative correlation has been observed between water hardness and death rate due to heart diseases (Masironi, 1979; Pocock et al., 1980; Teitge, 1990). Even though a definite causal effect still cannot be ascribed to this geochemical correlation, the effect of trace elements in drinking water on heart diseases has caused great interest among medical geologists. It is of particular interest to note that such a negative association between water hardness and cardiovascular pathology is evident in both industrialized and developing countries in the tropics.

It should, however, be mentioned that not all studies confirm such a relationship. Miyake and Iki (2004) for example, in a recent study, observed that there is a lack of association between water hardness and coronary heart diseases (CHD) mortality in Japan. These authors observed that in males, after adjustment for age, an inverse dose-response relationship between water hardness and mortality from CHD was significant (p=0.004). However, the relationship virtually disappeared after further adjusting for socioeconomics status and health care status. In females they found no association between water hardness and coronary mortality. Nonetheless, a large number of studies covering many countries suggest such a correlation and geochemically it is worthy of serious study.

WATER HARDNESS

Water hardness has been defined in the literature in a variety of ways with multiple units being used to express it, such as German, French and English degrees; equivalent $CaCO_3$ or CaO in mg/L. Even though initially water hardness was rather vaguely defined as a measure of the capacity of water to precipitate soap, it is now generally accepted that hardness is defined as the concentrations of calcium and magnesium ions or as $CaCO_3$ equivalent in mg/L. General guidelines for classification of water are given below:

$CaCO_3$ mg/L	Water hardness
0-60	Soft
61-20	Moderately hard
121-180	Hard
>180	Very hard

Most natural water supplies contain at least some hardness due to dissolved calcium and magnesium bearing carbonates and silicates. Elements such as iron may contribute to the hardness of water, but in natural water, they are generally present in low quantities. The total hardness of water may range from trace amounts to milligrams per litre.

Cardio-protective Role of Calcium and Magnesium

The *water factor* in the heavily discussed association of water hardness with the low incidence of cardiovascular diseases (CVD) has been the most intriguing question. What is it in the hard water that is really responsible for the cardio-protective role? Based on a large number of research investigations, the increasing evidence is indicative of magnesium as the *candidate element* with calcium playing a sub-ordinate supportive role (Eisenberg, 1992).

The presence of calcium and magnesium in natural water results from the decomposition of calcium and magnesium aluminosilicates and, at higher concentrations, from the dissolution of limestone, magnesium limestone, magnesite, gypsum and other minerals. In most waters, Ca and Mg are present as simple ions Ca^{2+} and Mg^{2+} with the Ca levels varying from tens to hundreds of mg/L and the Mg concentrations varying from units to tens of mg/L. The crustal abundance of Mg is much lower as compared to Ca and hence the lower abundance of Mg in the natural waters, the average Ca/Mg ratio being 4.

It is well known that both Ca and Mg are essential for the human body. Apart from being a major component of bones and teeth, they both play a role in the decrease of neuromuscular excitability, myocardial system, heart and muscle contractility, intracellular information, transmission and blood coagulability. Among the common manifestations of Ca-deficiency is osteoporosis and osteomalacia, while hypertension is also thought to be linked to Ca-deficiency.

Mg is a vital cofactor and activator of more than 300 enzymatic reactions including glycolysis, ATP metabolism, transport of elements such as Na, K and Ca through membranes, synthesis of proteins and nucleic acids, neuromuscular excitability, muscle contraction (Kožíšek, 2003). Mg-deficiency is known to be linked to e.g. vasoconstrictions, hypertension, cardiac arrhythmia, acute myocardial infarction.

An important point to note is that although only two out of three studies have shown correlation between cardiovascular mortality and water hardness (Figure 8.1), studies carried out on the water magnesium alone have all shown an inverse correlation between cardiovascular mortality and water magnesium level (Durlach et al., 1985).

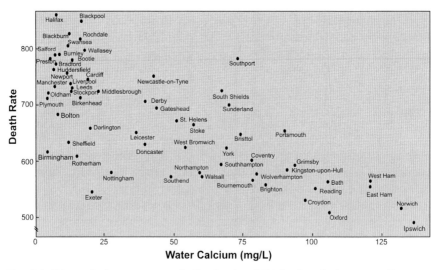

Fig. 8.1. Water Ca in some areas in England and Wales in relation to cardiovascular death rate (per 100,000 men), age 45-64 years during 1958-64 (Gardner 1973; Keil, 1979)

Neri et al. (1975, 1977) who carried out a nation-wide survey of 15 elements present in 575 drinking water samples were of the opinion that Mg was the element most-likely to show the cardioprotective role. They based their conclusions on the following evidence as reported by Marier (1978).

(i) It was present in more than 10% of sampled waters.

(ii) Mg has a consistent function of the softness-hardness gradient.

(iii) Mg represents a significantly high proportion of the daily intake from other sources.

(iv) The known metabolic effects of Mg are consistent with the hard water mortality trend.

(v) Analysis of 350 tissue samples from 161 autopsy cases revealed that myocardial Mg was 6% lower in *"cardiac death"* patients from soft-water localities in comparison with hard water regions.

(vi) Myocardial Mg in all *"cardiac death"* tissues averaged 22% lower than in the group of non-cardiac fatalities.

In the search for the "unknown water factor" associated with hard water, trace elements beneficial to health (Li, Zn, Co, Cu, Sn, Mn) and toxic (Pb, Cd, Hg) had also been investigated. No significant correlation between the content of these elements in water with CVD morbidity had been observed. Attention had been paid also to the higher corrosive potential of soft water known to support higher leakage of toxic compounds from water pipes.

Rubenowitz et al. (2000) studied the correlation between the drinking water Mg and Ca levels and morbidity and mortality from acute myocardial infarction (AMI) in 823 males and females aged 50-74 years in 18 Swedish districts. Their findings supported the hypothesis that Mg prevents sudden death from AMI, rather than all ischaemic disease deaths.

Maier (2003) observed that there is a link between low magnesium and atherosclerosis. It was shown that Mg-deficiency caused by poor diet and/or errors in its metabolism may be a missing link between diverse cardiovascular risk factors and atherosclerosis. The latter is described as a form of chronic inflammation resulting from interaction between lipoproteins, monocyte-derived macrophages, T cells and the normal components of the arterial wall.

Apart from Mg-deficiencies promoting inflammation, it is also known to be frequently associated with hypertension, diabetes and aging - known risk factors of atherosclerosis.

It has been shown (WHO, 1978; Oh et al., 1986), that soft water markedly reduces the content of different elements (including Mg and Ca) in food if used for cooking vegetables, meat and cereals. For Mg and Ca this figure could be as high as 60%. As against this if hard water is used for cooking, this loss was estimated to be much lower. Kožíšek (2003) was of the view that in areas supplied with soft water, one needs to take into account not only a lower intake of Mg and Ca from drinking water, but also a lower intake of Mg and Ca from food due to their loss during cooking in such water.

Several attempts have been made to quantify the protective effect of water magnesium (Figure 8.2). The study by Schroeder (1960) in USA estimated that an increase of the water Mg level by about 8 mg/L led to reduction of mortality from all CVD by about 10%. Teitge (1990), in his extensive

German study reported that if drinking water Mg is reduced by about 4.5 mg/L, the incidence of myocardial infarction increases by 10%.

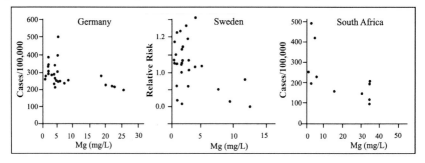

Fig. 8.2. Relation between cardiovascular death among men and drinking water magnesium levels in Sweden, Germany and South Africa (Rylander, 1996)

Even though the role of Mg in cardiovascular diseases has been more important, the Ca/Mg ratio has also been a subject of interest (Figure 8.3). However, there are no well defined values for such a ratio, bearing in mind, the still hypothetical nature of the importance of Ca and Mg in drinking water in CVD.

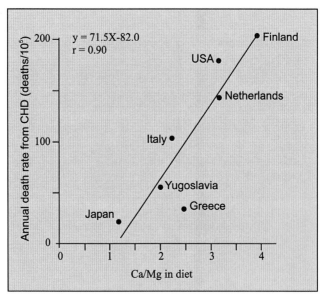

Fig. 8.3. Relationship between death rates from Coronary Heart Diseases (CHD) and the average calcium to magnesium ratio of the diet in different Organization for Economic Co-operation and Development (OECD) countries (Karppanen et al., 1978)

GEOCHEMICAL BASIS FOR TROPICAL ENDOMYOCARDIAL FIBROSIS (EMF)

Endomyocardial fibrosis (EMF) is an idiopathic disorder of people living in the tropical and subtropical regions of the world and characterized by the development of restrictive cardiomyopathy. It is sometimes considered to be a part of a spectrum of single disease processes that includes Löffler endocarditis (non-tropical eosiniophilic endomyocardial fibrosis). EMF displays a growth of a thick meshwork of fibrous tissues (collagen and elastin) within the endocardium and heart valves (Fergeson, 2002; Smith et al., 1998). EMF has been observed in Uganda, Nigeria and India among other tropical countries.

The aetiology of EMF is uncertain. One of the suggested reasons for tropical EMF is the influence of the element cerium (Eapen, 1998; Smith, 1998). The presence of elevated levels of dietary Ce and deficient levels of Mg, notably in South India have been considered as potential cofactors in the aetiology of endomyocardial fibrosis. Figure 8.4 shows the concentration of EMF in the equatorial region.

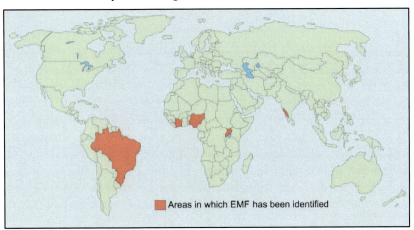

Fig. 8.4. World wide distribution of endomyocardial fibrosis (EMF) (Smith el al., 1998, Valiathan et al., 1993)

Valiathan et al. (1993) pointed out that the geochemistry of these tropical countries has common characteristics and it is the geochemistry that is perhaps responsible for tropical EMF. These authors studied the tropical EMF of Kerala in South India and made several observations. The majority of the tropical EMF cases were concentrated in the coastal areas where the Ce-rich mineral monazite was also concentrated (Fig 8.5). The monazite from the coastal sands from 2 sites of Kerala had cerium contents ranging from 34 to 37%. Cerium is the most abundant element in monazite and shows the highest solubility among all rare-earth metals. It is now known to be toxic to humans.

Valiathan et al. (1993) hypothesised that the non-random distribution of EMF in Kerala and its spatial coincidence with latosolic soils is indicative of geochemical factors having a causal association with the disease.

Kartha et al. (1993), in an attempt to develop an animal model for EMF based on the geochemical hypothesis, carried out experiments on rats using Mg and C. They claimed that there is evidence for the possible connection of EMF with myocardial levels of these two elements. The supportive data were the preferential accumulation of Ce in cardiac tissues compared to skeletal muscle, enhancement of cerium levels in tissues in Mg deficiency and the synergistic effect of Mg-deficiency, and the severity of myocardial lesions.

The presence of Th and Ce in conjunction with Mg-deficiency was indicative of the possibility that EMF could be the "cardiac expression of an elemental interaction that causes a toxic metal to replace an essential element at the cellular level" (Valiathan et al., 1993). An interesting observation made by these authors was that children suffering from helminthiasis and anaemia in Kerala were known to consume sand. The radioactivity of urine measured in children of Kerala between 5 and 9 was 21+2±7.1 pCi/L. They also refer to a case study in Minas Gerais, Brazil, where human ingestion of Th and REE based on their contents in faeces revealed a tandem relationship between Th and Ce (Linslata et al., 1986).

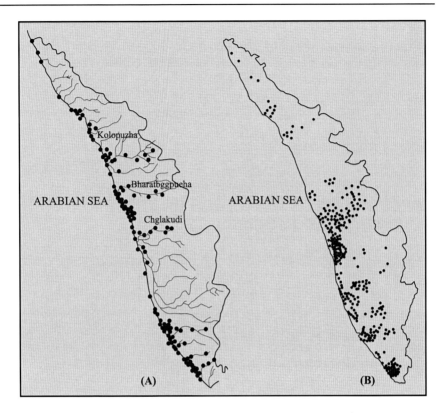

Fig. 8.5. (A) Distribution of monazite sand along Kerala coast, India and (B) Distribution of places of origin of 300 patients with endomyocardial fibrosis in (Valiathan et al., 1993)

The role of Mg was considered to be synergistic insofar as it enhances the adsorption of Ce and provides binding sites for the toxic element in the myocardium. If this geochemical hypothesis is correct, then similar over exposures to Ce must occur in other environments where EMF is prevalent. Smith et al. (1998) tested this hypothesis in Uganda where EMF is endemic and which represents the most common form of infantile heart condition. In Uganda, the presence of elevated Ce and lack of Mg in the surface environment was associated with the presence of highly weathered ancient granites and gneisses that form latosolic soils over large tracts of central and northern parts. Mineralogical and analytical examinations revealed the presence of a Ce-bearing non-phosphate mineral (with no significant La or P) possibly carbonate, oxide or hydroxide. Smith et al. (1998) were of the view that the occurrence of Ce without La and other REE indicates a mineral formed at low temperatures at the near-surface environment. Cerium present in the finer fractions of dust increases its mobility, bioavailability in the human gastrointestinal tract and the direct absorption through the skin (Price and Henderson, 1981) or scavenging cells within the gastrointestinal tract (Powell et al., 1996).

The work of Smith et al. (1998) substantiated the observations made in south-western India, and showed that Ce within the Ugandan environment is controlled by the presence of <20 µm particles in the soil.

EFFECT OF WATER HARDNESS ON URINARY STONE FORMATION (UROLITHIASIS)

The mineral and electrolyte content of drinking water varies markedly depending on the geology, soil chemistry and hydrology of the terrains concerned. The effect of these minerals and electrolytes on the human physiology has been the subject of many investigations. Urinary stone formation is common in many countries and particularly in some of the tropical countries (Singh et al., 2001). Figures 8.6 and 8.7 show some urinary stones and their internal structures.

Stones which form in the kidneys are made of different types of crystals.

There are made up of:

- (a) calcium oxalate
- (b) calcium phosphate
- (c) combination of calcium oxalate and calcium phosphate
- (d) magnesium ammonium phosphate (known as struvite or infection stones)
- (e) uric acid
- (f) cystine
- (g) miscellaneous types which may occur with drug metabolites.

Water Hardness

Fig. 8.6. A kidney stone with a diameter of about 6 cm

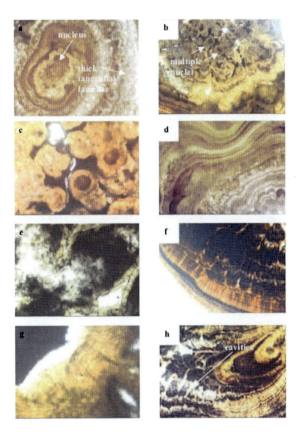

Fig. 8.7. Internal structures of kidney stones (Wijewardana, 2005)

Types of Stones

Calcium oxalate

This is the most common type of stone, and forms due to excessive amount of calcium oxalate in the urine. They are of two types namely (i) monohydrate (ii) dihydrate. While calcium oxalate dihydrate stones break easily with the treatment procedure lithotripsy, monohydrate stones are among the most difficult to break. These stones usually develop when the urine is acid (pH <6.0) (Leonard, 1961; Gibson, 1974; Kadurin, 1998; Sokol et al., 2005).

Calcium phosphate

Calcium phosphate develops in alkaline urine (pH >7.2) and often occur in people with urinary tract infection, hyperparathyroidism, medullary sponge kidneys or Renal Tubular Acidosis (RTA). Citrate is known to play an important role in calcium stone formation. It forms a soluble salt with calcium and inhibits the formation of calcium oxalate and calcium phosphate crystals. Low levels of urinary citrate therefore increase the chances of developing stones. However, when urine pH is less that 5.5, uric acid crystals develop and calcium crystals then form layers around the crystal to form a calcium oxalate stone.

Uric acid

In their pure form these stones do not contain calcium. Uric acid is an end product of urine metabolism and the crystals cause gout, an arthritic condition. In acid urine (pH <5.5) uric acid crystals precipitate leading to stone formation. When urine is alkaline, uric acid remains soluble.

Magnesium ammonium phosphate stones (struvite)

These stones are termed infection stones or struvite. They make up approximately 15% of the urine stones and are thus an important group. The basic precondition for the formation of infection stones is a urease positive urinary tract infection. Urease splits the urea to ammonia and CO_2. Alkaline urine also develops and struvite and carbonate apatite crystals develop (Bichler et al., 2002).

The following reactions take place in the formation of carbonate apatite and struvite in ureas positive infection. Urea is hydrolysed in the presence of urine.

$$H_2N-\underset{urea}{\underset{\|}{\overset{O}{C}}}-NH_2 \xrightarrow{urease\ +H_2O} CO_2 + NH_3$$

Ammonia and carbon dioxide hydrolyze to ammonium ions and bicarbonate. Binding with available cations produces magnesium ammonium phosphate (2) and carbonate apatite.

$$6\ H_2O + Mg^{2+} + NH_4^+ + PO_4^{3-} \underset{pH \leq 7.2}{\overset{pH \geq 7.2}{\rightleftarrows}} MgNH_4PO_4 \cdot 6H_2O \quad \text{Struvite}$$

$$CO_3^{2-} + 10\ Ca^{2+} + 6PO_4^{3-} \underset{pH \leq 6.8}{\overset{pH \geq 6.8}{\rightleftarrows}} Ca_{10}(PO_4)_6CO_3 \quad \text{Carbonate apatite}$$

Cysteine

These stones are different from the others since they do not have a protein matrix due to excess amounts of the amino acid cysteine in the urine (cystinuria). Due to the protein matrix, these stones are difficult to fragment.

The urinary stones, as discussed above are minerals and can be studied by techniques such as X-ray diffraction and optical microscopy. Among these minerals are:

(a) Apatite	$Ca_5(PO_4,CO_3)_3\ (F,OH,Cl)$	calcium phosphate
(b) Whewellite	$CaC_2O_4 \cdot H_2O$	calcium oxalate
(c) Weddellite	$CaC_2O_4 \cdot 2H_2O$	
(d) Struvite	$Mg(NH_4)PO_4 \cdot 6H_2O$	magnesium ammonium phosphate
(e) Brushite	$CaHPO_4 \cdot 2H_2O$	calcium hydrogen phosphate
(f) Whitlockite	$Ca_9(Mg, Fe)H(PO_4)_7$	calcium phosphate (more common as prostate stone)
(g) Newberyite	$MgHPO_4 \cdot 3H_2O$	magnesium hydrogen phosphate

Kajander and Ciftcioglu (1998) in a controversial theory attributed stone formation and calcification in the human body to nanobacteria. These authors found nanobacterial culture systems that allow for reproducible production of apatite calcifications in vitro. Depending on the culture conditions, tiny nanocolloid-sized particles covered with apatite, forming various sizes of aggregates and stones were observed. They considered nanobacteria as important nidi (microcrystalline centres) for crystal formation. They concluded that bacteria- mediated apatite takes place in humans just as in the case of aqueous environments and geological materials, notably sediments. This theory however was criticized by Cisar et al. (2000) who proposed that biomineralization was caused by the non-living, nucleating activities of self- propagating microcrystalline centres (nidi) which form crystalloid macromolecules of calcium carbonate phosphate apatite. Photographs of these non-living structures taken by electron microscopy were identical in appearance to those previously described by Kajander and Ciftcioglu (1998).

Studies carried out on the effect of water hardness on urinary stone formation (Schwartz et al., 2002) appear to indicate a lack of any significant association between water hardness and the incidence of urinary formation. Bellizia et al. (1999) in their studies on effects of water hardness on urinary risk factors for kidney stones in patients with idiopathic nephrolithiasis observed that, as compared with both tap and soft water, hard water was associated with a significant 50% increase of the urinary calcium concentration in the absence of changes of oxalate excretion.

Many early investigators have documented an inverse relationship between drinking water hardness and calcium urolithiasis (Churchill et al., 1978; Churchill et al., 1980; Shuster et al., 1982). The reason for such an inverse relationship was not well known. As pointed out by Schwartz et al. (2002), increased mineral content, most notably calcium, in hard water is the biggest concern of most patients. However, these authors contend that no studies support the premise that ingesting hard water increases the risk of urinary stone formation. Singh et al. (2001) studied the role of fluoride in urinary stone formation in humans using case studies in India. They selected two areas, a fluoride endemic area and a fluoride non-endemic area. They observed that the prevalence of urolithiasis was 4.6 times higher in the endemic area as compared to the non-endemic area. Furthermore, they noted that the prevalence was almost double in subjects with fluorosis as against those without fluorosis in the endemic area. However, the effect of fluoride in drinking water on urinary stone formation is still not sufficiently well known.

CHAPTER 9

SELENIUM- A NEW ENTRANT TO MEDICAL GEOLOGY

The element selenium could be regarded as a new entrant to the field of medical geology. It is of special importance to the developing countries of the tropical belt where there appears to be an apparent association with some diseases. Since selenium is not an element that is determined in routine investigations, the real impact of selenium excess or deficiency on the population may be hidden. Recently however, the importance of selenium as a medically important element has received increasing attention and many studies, notably in China, have been carried out on the epidemiology of selenium-associated diseases.

Selenium was designated as an essential trace element for humans and animals in the late 1950s. It is a part of the biologically important enzyme glutathione peroxidase (GSH-PX) which acts as an antioxidant preventing tissues degeneration. However, as for some other trace elements, excessive doses could cause ill health. The range between deficiency levels, (<11 µg/g per day) and toxic levels in susceptible people (>900 µg/g per day) is very narrow (Fordyce et al., 2000b; Yang and Xia, 1995). The endemic degenerative heart disease in China, known as Keshan Disease and an endemic osteoarthropathy (Kaschin-Beck Disease) causing deformity of affected joints, are attributed to selenium deficiency.

THE GEOCHEMISTRY OF SELENIUM IN THE ENVIRONMENT

Selenium is classified as a metalloid -an element which has properties of both a metal and a non-metal. It shows chemical similarities to sulphur, which also lies in Group VI of the periodic table. Because of this similarity in form and components they have many interrelations in biology (Adriano, 2001). Selenium exhibits a number of oxidation states, -II, 0, +IV and +VI. The most important oxidation states of selenium in the environment are -II (selenides) 0 (ground state) +IV (selenites) and +VI (selenates). The abundance of Se in the earth's crust ranges from 0.05 to 0.09 mg/kg, approximately 1/6000th of sulphur and 1/50th of that of arsenic. Table 9.1 shows the selenium concentrations in some environmental media. Coal and black shales are generally enriched in selenium. Approximately 50 selenium minerals are known and on account of the similarity of Se with S, it is commonly associated with heavy metal (e.g. Ag, Cu, Pb, Hg, Ni) sulphides and occurs as either a selenide or as a substitute ion for S in the crystal lattice (Malisa, 2001).

The mobility and bioavailability of selenium is determined by the chemical form. Table 9.2 shows the common chemical forms of selenium in geological and biological materials. The general geochemistry of selenium is complex since it shows several oxidation states and it has the ability to complex with organic matter (Figure 9.1). Selenium is therefore found in a variety of naturally occurring materials, the main sources being rock weathering, volcanic emissions and metal sulphide deposits.

In sedimentary rocks, Se is associated with the clay size fraction and hence it is more abundant in shales as compared to limestones or sandstones. It is also known that relatively high concentrations (>300 µg/g) of Se are found in some phosphate rocks and in view of the fact that phosphate fertilizers are commonly used, this may be an important source of selenium in the environment (Frankenberger and Engberg, 1998).

The Se concentrations of most soils range from 0.01-2 µg/g (world mean 0.4 µg/g; Fergusson, 1990). In some seleniferous areas, however, Se concentrations up to 1200 µg/g have been reported (Mayland et al., 1989). As shown in Figure 9.2, in acid and neutral soils inorganic Se occurs as very insoluble Se^{4+} compounds of oxides and oxyhydroxides of ferric iron. In neutral and alkaline soils, Se^{6+} is the main oxidation state, being soluble and hence more bioavailable. A notable geochemical feature is that Se^{4+} is adsorbed onto soil particle surfaces, with greater affinity than Se^{6+} (Fugita et al., 2005). The soil geochemistry of selenium is controlled by many factors-among which are:

       (a) Se-speciation
       (b) soil texture
       (c) mineralogy
       (d) organic matter
       (e) presence of competing ions
       (f) iron-oxides
       (g) clays.

Table 9.1. Selenium concentrations in selected materials (McNeal and Balistrieri, 1989; Fordyce et al., 1998; Malisa, 2001).

Material	Se (µg/g)
Earth's Crust	0.05
Granite	0.01-0.05
Limestone	0.08
Sandstone	<0.05
Shale	0.06
Phosphate rock	1-300
Granite	0.025
Soil	
USA	<0.1-4.3
England/Wales	<0.01-4.7
Seleniferous	1-80, <1200
Coal	0.46-10.65
Atmospheric dust	0.05-10
River water	
Mississippi	0.00014
Amazon	0.00021
Colorado (alkaline)	0.01-0.4
Lake Michigan	0.0008-0.01
Sea water	0.000009
USA plants	
Grasses	0.01-0.04
Clover and alfalfa	0.03-0.88
Barley	0.2-1.8
Oats	0.15-1
Algae	
Marine	0.04-0.24
Freshwater	<2
Whole fish	
Marine	0.3-2
Freshwater	0.42-0.64
Animal tissue	0.4-4

Table 9.2: Common chemical forms of Se in the environment.

Oxidation state	Chemical forms
Se⁰	Elemental selenium
Se⁴⁺	Selenite SeO$_3^{2-}$
	Trimethylselenonium (TMSe) (CH$_3$)$_3$Se$^+$
	Selenous acid H$_2$Se$_3^{3-}$
	Selenium dioxide SeO$_2$
Se⁶⁺	Selenate SeO$_4^{2-}$
	Selenic acid H$_2$SeO$_4^{2-}$
Se²⁻	Selenides Se^{2-}
	Dimethylselenide (DMSe)(CH$_3$)Se
	Dimethyldiselenide (DMdSe)(CH$_3$)$_2$Se$_2$
	Hydrogen selenide H$_2$Se
	Dimethylselenone (CH$_3$)$_2$SeO$_2$

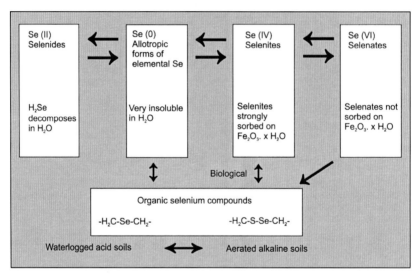

Fig. 9.1. Generalized chemistry of selenium in soils and weathering sediments (Allaway, 1968)

The bioavailability of selenium is particularly influenced by clay and organic matter and these, due to their capacity to trap selenium, lower the bioavailability. The presence of SO$_4^{2-}$ and PO$_4^{3-}$ which compete with Se ions for fixation sites in soils and plants are also known to affect the bioavailability of selenium (Fordyce et al., 1998).

The Se concentrations in water, are generally very low and only rarely exceed the WHO safely limit of 10 µg/L. MacGregor (1998), studied a number of wells in the Amman-Zarqa Basin in Jordan and reported Se concentration up to 12 times the WHO safety range of values being 0.8 µg/L to 112 µg/L. The wells with highest levels were situated extremely close to an abandoned phosphate mine.

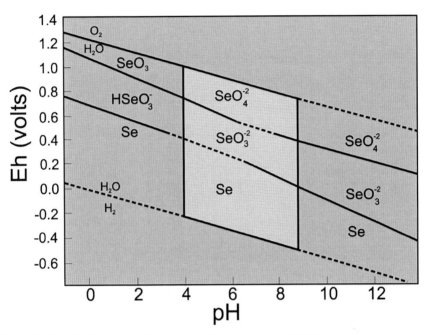

Fig. 9.2. Eh-pH diagram of Se in soils (Mayland et al., 1989)

Selenium is considered non-essential for plants. However, certain plants, termed hyper-accumulators have the ability to absorb and accumulate high concentrations of selenium. Plants absorb selenium in the form of selenate, selenite or organic-Se. Selenate is absorbed by the root system through the binding sites of sulphate. In view of the similarities, S and Se show similar biochemical reactions, though in competition with each other. Selenite, however, is thought to be taken up through different sites. The biochemical similarity of Se and S are seen in their metabolism by the same enzymes and assimilatory pathways (Adriano, 2001). The formation of Se analogues of S compounds that are substrates for S assimilation enzymes therefore, cause Se- toxicity. This is mainly due to the likely replacement of S by Se in the amino acids of proteins, which in turn disrupts the essential catalytic reactions (Figure 9.3).

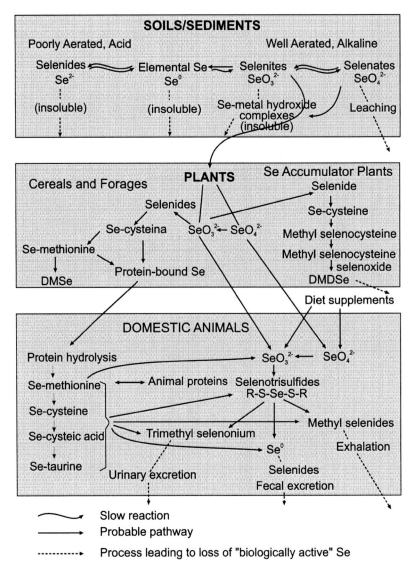

Fig. 9.3. Biogeochemical reactions and pathways of selenium in the soil-plant-animal system (Adriano, 2001)

The hyperaccumulation of selenium by some plants, which in some cases rises to more than 0.5 wt.%, is of special interest. As shown in Figure 9.4, while normal plants (non-adapted) do not separate S from Se biochemically, the hyper-accumulators are able to separate inorganic sulphur (as sulphate) from inorganic selenium (as selenate or selenite) as they enter the plant. Hence Se toxicity is avoided.

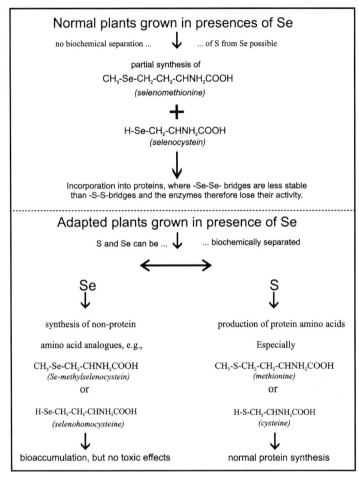

Fig. 9.4. Adaptation and bioaccumulation of selenium in plants (Streit and Stumm, 1993)

Microbial Transformation of Selenium

The oxidation of selenium and its subsequent geochemical pathways in the environment involve to a large extent microbial reduction, oxidation, methylation and demethylation reactions. Biological transformations of toxic Se-oxyanions into less toxic or biologically unavailable forms, such as Se or volatile Se compounds are being investigated thoroughly in view of their potential use in bioremediation (Dungan and Frankenberger, 1999).

Since both selenate [SeO$_4^{2-}$; Se(VI)] and selenite [SeO$_3^{2-}$;Se (IV)] are toxic and show bioaccumulation, the removal or immobilization of these oxyanions has received greater attention.

The selenium cycle in soil is illustrated in Figure 9.5. As in the case of sulphur, microbial transformations predominate in the cycling of selenium. There are four biological transformations that are of importance. These are (i) reduction (assimilatory and dissimilatory) (ii) oxidation (iii) methylation (iv) demethylation.

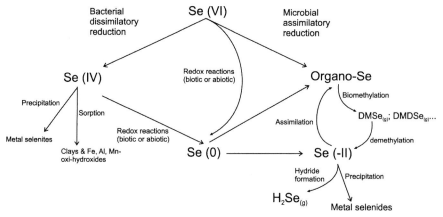

Fig.9.5. Mobility of selenium in the environment

Assimilatory reduction is the reduction and incorporation of selenium into organic compounds and dissimilatory reduction is when microorganisms reduce SeO$_4^{2-}$ as a terminal electron acceptor in energy metabolism. As noted by Dungan and Frankenberger (1999), the focus of attention has been on the dissimilatory reduction of Se-oxyanions in view of their potential application in remediating seleniferous environments by producing the biologically unavailable Se^0. Methylation of Se is considered to be a mechanism used by microorganisms to protect themselves against Se-toxicity in Se-rich environments. The volatilization of Se which follows the methylation reactions then removes the selenium from the toxic environment. Tables 9.3 and 9.4 list the selenium reducing microorganisms.

Dissimilatory Reduction

The reduction of both Se-oxyanions to Se^0 in the soil-sediment-water systems is caused by several types of bacteria, the majority of them being SeO$_3^{2-}$ reducers (Bautista and Alexander, 1972; Doran, 1982). It has been demonstrated by Oremland et al. (1989) that the bacterial reduction of SeO_4^{2-} to the non-bioavailable Se^0 is a major sink for Se-oxyanions in anoxic sediments.

Table 9.3. Selenium reducing microorganisms (Dungan and Frankenberger, 1999)

Organisms	Description of Reaction
Enterobacter cloacae SLD1a-1	Respires SeO_4^{2-} and NO_3^-, and reduces SeO_4^{2-} to Se^0 only in the presence of NO_3^-
Thauera selenatis	Grows anaerobically using SeO_4^{2-}, NO_3^-, and NO_2^-. Reduction of SeO_4^{2-} occurs by way of a SeO_4^{2-} reductase. SeO_4^{2-} is completely reduced to SeO only when NO_3^- is present
Strain SES-3	Respires SeO_4^{2-} and can reduce SeO_3^{2-} to Se^0 in washed-cell suspensions
Pseudomonas stutzeri	Reduction of SeO_4^{2-} and SeO_3^{2-} to Se^0 under anaerobic conditions
Wolinella succinogenes	Adapted cultures able to reduce SeO_4^{2-} and SeO_3^{2-} to Se^0 under anaerobic conditions
Desulfovibrio desulfuricans	SeO_4^{2-} and SeO_3^{2-} were reduced to Se^0 under anaerobic conditions, but both Se oxyanions could not support growth
Salmonella heidelberg	SeO_3^{2-} reduced aerobically to Se^0
Streptococcus faecalis N83 *Streptococcus faecium* K6A	Anaerobic reduction of SeO_3^{2-} to Se^0 by resting-cell suspensions
Clostridium pasteurianum	Reduction of SeO_3^{2-} by hydrogenase (I)
Bacillus subtilis *Pseudomonas fluorescens*	Reduction of SeO_3^{2-} to Se(0) by a NO_2^- and SO_3^{2-} independent enzyme system

Assimilatory Reduction

Both selenate and selenide undergo assimilatory reduction to Se^{2-} which gets incorporated into cellular proteins, similar to the mechanism of sulphur incorporation in amino acids. The products, the selenoamino acids such as selenomethionine and selenocysteine have been produced by some bacteria and yeast. However with excess Se toxicity effects appear and the organisms undergo metabolic deterioration.

Table 9.4. Selenium volatilizing microorganisms (Dungan and Frankenberger, 1999)

Organisms	Se Substrate	Se Product
Fungi		
Scopulariopsis brevicaulis	SeO_4^{2-}, SeO_3^{2-}	DMSe
Penicillium notatum	SeO_4^{2-}, SeO_3^{2-}	DMSe
Penicillium chrysogenum		
Schizopyllum commune	SeO_3^{2-}	DMSe
Aspergillus niger	SeO_4^{2-}	DMSe
Penicillium sp.	SeO_3^{2-}	DMSe
Penicillium sp.	SeO_4^{2-}, SeO_3^{2-}	DMSe
Fusarium sp.		
Cephalosporium sp.		
Scopulariopsis sp.		
Candida humicola	SeO_4^{2-}, SeO_3^{2-}	DMSe
Alternaria alternata	SeO_4^{2-}, SeO_3^{2-}	DMSe
Penicillium citrinum	SeO_3^{2-}-	DMSe, DMDSe
Acremonium falciforme		
Penicillium sp.	SeO_3^{2-}	Unidentified
Bacteria		
Corynebacterium sp.	SeO_4^{2-}, SeO_3^{2-}, Se^0	DMSe
Aeromonas sp.	SeO_3^{2-}	DMSe, DMDSe
Flavobacterium sp.		
Pseudomonas sp.		
Pseudomonas fluorescens K27	SeO_4^{2-}	DMSe, DMDSe, DMSeS
Rhodocyclus tenuis	SeO_4^{2-}, SeO_3^{2-}-	DMSe, DMDSe
Rhodospirillum rubrum S1		
Aeromonas veronii	SeO_4^{2-}, SeO_3^{2-}, Se^0, SeS_2, H_2SeO_3, SeH	DMSe, DMDSe, methylselenol, DMSeS

Oxidation

The oxidation of reduced forms of Se to Se-oxyanions is an important reaction in the environment not only because SeO_4^{2-} and SeO_3^{2-} are soluble and toxic, but also because biomethylation of Se depends mainly on the oxidized forms of Se (Dungan and Frankenberge, 1999). In view of the similar biochemistry of S and Se, it is expected that microbial oxidation of Se occurs in a manner similar to sulphur, both heterotrophic and autotrophic organisms being able to carry out the reaction.

Methylation and Volatilization

In soils, water and sediment enriched in selenium, methylation of Se-compounds from Se-oxyanions and organo-Se compounds can take place (Doran, 1982). This is essentially a mechanism of detoxification of the environment by some bacteria and fungi in soils and by bacteria in the aqueous systems. The methylation of selenite to dimethyl selenide by the soil bacterium Corynebacterium, was proposed by Doran (1982) to follow the mechanism.

$SeO_3^{2-} \rightarrow Se^0 \rightarrow HSeX \rightarrow CH_3SeH \rightarrow (CH_3)_2Se$

Selenite   elemental Se   selenide   monomethanol   dimethyl-
                                                          seleniumhydride   selenide

For the aqueous system, Cooke and Bruland (1987) proposed a pathway for the reduction of dimethylselenide.

$SeO_3^{2-} \xrightarrow{\text{assimilation and reduction to (-2)}} CH_3Se(CH_2)_2CHNH_2COOH$ selenomethionine

$CH_3Se(CH_2)_2CHNH_2COOHCH_3 \xrightarrow{\text{methylation}} (CH_3)_2Se^+(CH_2)_2CHNH_2COOH$
dimethyl selenomethionine
(dimethyl selenonium ion)

$(CH_3)_2Se^+(CH_2)_2CHNH_2COOH \xrightarrow{H_2O} CH_3SeCH_3 + CH_3COHCH_2CHNH_2COOH$
                                                          dimethylselenide   homoserine

Methyl cobalamin and S-adenosylmethionine are known as the methyl donors in the microbial methylation of Se. Dimethylselenide is considered to be the major volatile species produced by most microorganisms.

SELENIUM AND HUMAN AND ANIMAL HEALTH

Recent research has shown that the trace element selenium is an essential nutrient and is of fundamental importance to human and animal health. Schwarz and Foltz (1957) first showed the essentiality of selenium when they demonstrated their prevention of liver necrosis in rats by selenium. Subsequently Thompson and Scott (1969) showed that poor growth and higher mortality rates of chicks was caused by Se-poor diets.

More recently (Deverel et al., 1984) selenium poisoning was found to be the major cause of high mortality of wildfowl at the Kesterson Wildlife Refuge in the San Joaquin Valley, California. Selenium from the selenium-rich country rock had been geochemically mobilized and found its way into the Refuge reservoir which was a nesting area for wildfowl. Both fish and wildfowl had very high levels of selenium.

Among the diseases associated with a low intake of Se are cardiomyopathy (Keshan disease), deforming arthritis (Kashin-Beck disease), protein energy malnutrition, haemolytic anaemia, hypertension, ischaemic heart disease, cancer, multiple sclerosis, muscular dystrophy, infertility, cystic fibrosis and alcoholic cirrhosis (Haygarth, 1994; Rotruer et al., 1993).

These case studies clearly demonstrate the essentiality and toxicity of selenium which shows a very narrow threshold window (>900 µg/g per day-toxicity and <11 µg/g per day). Selenium is a component of selenoproteins some of which have important enzymatic functions. Selenocystine is the 21st aminoacid. It has been now recognized that all these enzymes are selenium-dependent, with selenocysteine as the active site (Sunde, 1997). The importance of selenium here is that it functions as a redox centre. Selenium is an important component of glutathione peroxidase (GSHPX), considered as a critical enzyme which prevents oxidative damage at the cellular level. It helps to maintain membrane integrity, protects prostacyclin production and reduces the likelihood of propagation of further oxidative damage to biomolecules such as lipids, lipoproteins and DNA with the associated increased risk of conditions such as atherosclerosis and cancer (Diplock, 1994; Néve, 1996; Rayman, 2000). Selenium deficiency diseases in animals are mostly seen in livestock and include reproductive impairment, growth depression and white-muscle disease, a myopathy of heart and skeletal muscle affecting mainly lambs and calves (Reilly, 1996). Rayman (2000) reviewed the importance of selenium to human health and discussed the health effects of less overt selenium deficiency. These deficiency-related health effects can be summarized as follows.

Immune function

Deficiency of selenium is thought to be accompanied by loss of immunocompetence. This is perhaps linked to the fact that selenium is normally found in significant amounts in immune tissues such as liver, spleen and lymph nodes. Selenium supplementation is also known to have marked immunostimulating effects.

Viral infections - AIDS

Selenium deficiency is associated with the occurrence, virulence or disease progression of some viral infections. It is a potent inhibitor of HIV replication in vitro. Selenium-deficient HIV patients are nearly 20 times more likely ($p < 0.0001$) to die from HIV-related causes than with sufficient levels (Baum et al., 1997). These authors defined selenium-deficiency as plasma concentrations at or below 85 µg/L. In the case of HIV-infected children, the low levels of plasma selenium were significantly and independently related to mortality (relative risk 5.96; $p = 0.02$) and faster disease progression (Campa et al., 1999).

Reproduction

Selenium is considered essential for male fertility. It is required for testosterone biosynthesis and the formation of and normal development of spermatozoa.

Mood

Rayman (2000) reports some studies where low selenium status had been linked to greater incidence of depression, anxiety, confusion and hostility.

Thyroid function

When selenium deficiency combines with iodine deficiency, hypothyroidism is enhanced and this is manifested as myxoedematous cretinism. This disease is known in the Democratic Republic of Congo (Zaire) where both iodine and selenium are known to be deficient (Vanderpas et al., 1990).

Cardiovascular diseases

Though not proven to any degree of certainty, selenium is thought to be protective against cardiovascular disease (Nève, 1996). Glutathione peroxidase, of which selenium is a component, is known to reduce hydroperoxides of phospholipids and cholesteryl esters associated with lipoproteins. This is expected to reduce the accumulation of oxidized low – density lipoproteins in the artery wall. The status of other antioxidants such as vitamin E however may complicate direct correlations.

Oxidative-stress or inflammatory conditions

Selenium is known to exhibit properties of an antioxidant and anti- inflammatory agent. In the case of pancreatitis, asthma and rheumatoid arthritis, selenium levels are considered to be important. In a study in Germany, intravenous administration of selenium to patients with necrotising pancreatitis reduced mortality from 89% in controls to zero in the treatment group (Kuklinsky and Schweder, 1996).

Cancer

An inverse relation between selenium intake and cancer mortality has been proposed in several epidemiological studies. The dietary intake of selenium in 27 countries was studied by Schrauzer et al. (1977). In the Harvard based Health Professionals' Cohort study (Yoshizawa et al., 1998) which involved the investigation of selenium intake and prostate cancer in 34000 men, it was observed that those in the lowest quintile of selenium status had 3 times the chances of developing cancer as against the highest quintile. It should however be emphasised again that correlation does not necessarily mean causation and further detailed studies are very necessary to attribute causal effects of selenium deficiency or toxicity to health.

SELENIUM DEFECIENCY DISEASES IN CHINA

The two diseases, Keshans's Disease (KD) and Kashin-Beck Disease (KBD) are particularly prevalent in some parts of China, and given China's population of over 1.3 billion(~20% global population) global Se-deficiency in China is of special interest in medical geology. Apart from KD and KBD, the impact of selenium-deficiency on human longevity is also an interesting study in China (Moffat, 1990; Foster and Zhang, 1995) bearing in mind that there is a relative lack of migration of people from village to village and the reliance of locally grown food which is influenced heavily by the soil chemistry.

Figure 9.6 illustrates the distribution of KSD and KBD in China. The low Se-belt (Figure 9.7) also known as the 'disease belt' stretches from Heilongjiang Province in the NE to Yunnan Province on the SW. The average abundance of Se in the earth's crust of China is 0.058 mg/kg and is lower than that in other parts of the world (Xia and Tang, 1990). In the different rock types, Se decreased in the order slate > clay rocks > basic and ultrabasic rocks > alkaline rocks > basalt > granite > hypersthene sandstones > limestones.

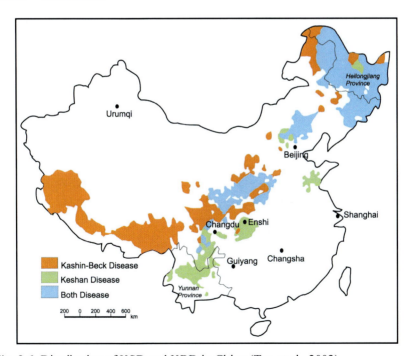

Fig. 9.6. Distribution of KSD and KBD in China (Tan et al., 2002)

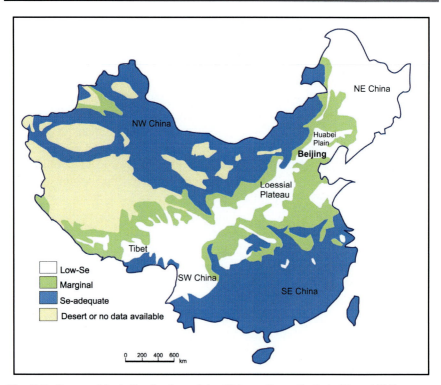

Fig. 9.7. Geographical distribution of the Chinese Low-Se Belt (Tan, 1989)

Wang and Gao (2001) report the occurrence of two notable Se-excessive or seleniferous regions in China, namely the Exi region in Hubei Province and the Ziyang region in Shanxi Province. The bedrock in the Exi seleniferous region consisted of carbonaceous shale and high Se coal.

Whereas the concentration of Se in fresh water around the world is 0.2 µg/L in the KD and KBD areas in China, in the low Se belt, Se concentration in well water is sometimes as low as 0.11 µg/L. In most Chinese cities and urban areas it is 0.65 µg/L. On the other hand, in the selenium-rich areas, two well water samples had 8.4 and 72 mg/L selenium (Wang, 1991).

Tan et al. (2002) studied the selenium contents in soil in the endemic disease areas in China and noted that the concentration of selenium in soil mainly depends on two groups of factors. These were (i) geographically azonal factors such as parent rocks and landforms which determine the source of Se in soil (ii) zonal factors such as biological and climatic factors, which influences the migration, maintenance, existing forms and availability of Se in e.g. soils, plants.

As shown in Table 9.5 in the KD and KBD areas, the concentrations of total-Se and water soluble Se both in cultivated and natural soils are lower than that in non-affected areas. The natural soils in the affected areas included dark brown soil, brown earth, drab soil, yellow-brown soil, red drab soil, loessial soil, purple soil and black soil. In disease-free areas, laterites, red soils, yellow soils, desert soils, chernozem, chestnut soil and calcic brown soils were found.

Table 9.5. Se contents of soils in the areas with and without KSD/KBD (Note: N: number of samples; X: arithmetic mean; G: geometric mean; S.D.: arithmetic standard deviation; S_{lg}: geometric standard deviation) (Tan et al., 2002)

Type	Area with KBD / KSD					Area without KBD / KSD				
	X	SD	G	S_{lg}	N	X	SD	G	S_{lg}	N
Total Se (mg/kg)										
Cultivated	0.112	0.057	0.100	0.1816	35	0.224	0.134	0.219	0.3297	161
Natural	0.119	0.075	0.105	0.2177	69	0.227	0.141	0.211	0.3192	86
Water soluble Se (µg/kg)										
Cultivated	2.5	1.0	2.5	0.3914	25	6.8	9.1	4.7	0.3106	151
Natural	2.8	2.2	2.2	0.2651	22	6.7	13.2	4.7	0.4452	71

Tan et al. (2002) concluded that the distribution of Se in top soil in China and its relationship to Keshan Disease and Kashin-Beck Disease is indicative of <0.125 mg/kg total selenium. The Se content in seleniferous areas was >3 mg/kg.

Fordyce et al. (2000b) studied the soil, grain and water chemistry in relation to human selenium responsive diseases in Enshi District China and noted that the majority of samples in the low selenium villages are deficient or marginal in Se and that Se availability to plants is inhibited by adsorption onto organic matter and Fe oxyhydroxides in soil. They also observed that in high Se-toxicity villages, the Se-bioavailability is controlled by the total soil Se concentration and pH. Johnson et al. (2000) were of the view that the organic content of the soils is a major factor in controlling the availability of Se and that the high incidence KD villages have the most organic-rich soils. Although higher in total Se, the organic-rich soils had little bioavailable Se resulting in a Se-deficient food chain.

SELENIUM AND IODINE DEFICIENCY DISEASES (IDD)

Recent research has shown that Se deficiency may be an important factor in the onset of IDD. It is known that the selenoenzyme, type-I iodothyronine deiodinase (IDI), is responsible for the conversion of the prohormone T_4 to the active hormone T_3 which exerts a major influence on cellular differentiation, growth and development, especially in the foetus, neonate and child. The conversion of T_4 to T_3 is inhibited by Se-deficiency and this affects thyroid hormone metabolism (Arthur and Beckett, 1994; Arthur et al., 1999).

Fordyce et al. (2000a) studied the relationship of IDD in Sri Lanka and Se deficiency and observed that it is unlikely that Se deficiency is the main controlling factor in IDD. However they were of the opinion that it could contribute to the onset of goitre along with iodine deficiency and other factors such as poor nutritional status and the presence of goitrogenic substances in the diet.

CHAPTER 10

GEOLOGICAL BASIS OF PODOCONIOSIS, GEOPHAGY AND OTHER DISEASES

The direct entry of soil into the human body, its effects, both biochemical and physiological, is an interesting field of study within medical geology. These processes enable geochemists and human and animal physiologists to study the direct link between soil chemistry, soil mineralogy and health. This chapter deals with two processes-geophagia and podoconiosis that clearly establish such a link. While the health effects of trace elements that enter the human body via food and water have been studied for many decades, the effect of inorganic minerals such as those found in rocks and minerals, on human and animal health is less known and has now aroused considerable scientific interest.

GEOPHAGY

Geophagy is defined as deliberate and regular consumption of earthy materials such as soils, clays and related mineral substances by humans and animals (Abrahams and Parson, 1996). Even though it is a practice that is found in all continents, it is most commonly seen in the tropics, notably in tropical Africa. This habit of soil-eating is particularly common among pregnant women and is listed as the 'craving' for extra-ordinary food or non-food substances (Geissler et al., 1999). Alexander von Humboldt, who explored South America observed the practice of geophagy during his expeditions in Orinoco in Venezuela during the period 1799-1804 (Keay, 1993). The Ottomac people who practiced this soil-eating habit in Venezuela apparently did not eat every type of clay but chose only the clays which contained the 'most unctuous earth and smoothest to touch'. Further, it had not caused any problems of health to these people, whereas other tribes who ate different soils became sick. This early observation by Alexander von Humboldt caused much interest in medical science much later and the debate whether there are district benefits in soil-eating, such as trace element inputs, and whether these geological materials bestow other physiological and biochemical benefits still goes on.

Wilson (2003) notes that many hypotheses have been advanced to explain geophagic behaviour, notably, detoxification of noxious or unpalatable compounds present in the diet, alleviation of gastrointestinal upsets such as diarrhoea, supplementation of mineral nutrients, and as a means of dealing with excess acidity in the digestive tract. Soil-eating is not only practiced by humans but animals such as monkeys, chimpanzees, gorillas, birds, reptiles and horses. Why do humans and animals consume soil? This intriguing question will probably be answered by a number of investigations currently being carried out to resolve conflicting views and to make an objective judgment regarding clinical, medicinal and nutritional implications of the practice (Reilly and Henry, 2000).

In the search for the geochemical characteristics that may confer medical benefits, clays figure most prominently in all the geophagic materials. Clay mineralogy is therefore considered to be an important component in the assessment of the different hypotheses.

One of the scientific approaches to test the possible beneficial effects of geophagic materials on human health is to first study their detailed chemistry and mineralogy and then to consider the biological effects of the elements and minerals thus isolated. Mahaney et al. (2000) studied the mineral and chemical analyses of soils eaten by humans in Indonesia. Five Javanese soil samples, including three earths eaten by humans as a therapeutic medicine, were analyzed along with the suitable control samples-those that were not eaten. The soils that were eaten had a high content of hydrated halloysite and kaolinite. The expandable clay mineral smectite was also present along with hydrated halloysite in a ratio of nearly 1.1. Mahaney et al. (2000) considered only Na, Mn, K, and S as the possible candidates that stimulate geophagy. What is worthy of note however, is that in all cases the eaten soils had predominantly higher levels of 1.1 clay minerals than the 2.1 minerals which predominated in control soils. These authors were of the view that soils can adsorb dietary toxins, normally present in the plant diet or those produced by microorganisms. The toxic alkaloids such as quinine, atropine and lupamine were thought to be adsorbed by these soils from Java and hence the soil-eating habit by some Javanese people.

According to a study by Johns (1986), some Indian tribes from Bolivia Peru and Arizona consumed four different geophagic clays of which the most widely used, was the interstratified illite-smectite together with smaller amounts of other clay minerals such as kaolin and chlorite. The explanation given was that geophagy functions as a detoxification method to enable the consumption of species of wild potatoes. Consuming clay with potatoes was considered to be effective in eliminating the bitterness of food and in preventing stomach pains and nausea. Further, all clays were effective in the adsorption of the glycoalkaloid tomatine over a range of simulated physiological conditions. Tomatine adsorption capacities of the clays were usually intermediate between those of pure smectite and kaolinite, but for one clay it was significantly higher than smectite (Wilson, 2003).

In many countries in Africa, kaolinite was the dominant geophagic clay which was even available in local markets (Johns and Duquette, 1991). Here the explanation for the geophagy was that it alleviates diarrhoea and sickness during pregnancy (Figure 10.1). Interestingly, this could be compared to the kaolinite-based western medicines taken for alleviating gastrointestinal malfunctions.

Fig. 10.1. "Clay Sweets" for pregnant women in Bangladesh (photo Dr. Rohana Chandrajith)

Mineral nutrient supplementation is one of the main explanations for the prevalence of geophagy among people living in tropical countries. This hypothesis was tested by Abrahams and Parsons (1997) who analyzed the chemical composition of geophagical soils from Thailand, Uganda and Zaire. It was reported that they all had low organic matter (0.2-1.5%) and clay contents (15-28%). Wilson (2003) however suggested that some soils had high cation exchange capacities (CECs) and the dominance of the non-clay fraction of the soils were indicative of inadequate dispersion, a common problem in highly weathered tropical soils. The sand and silt fractions also were considered to contain clay minerals in an aggregated form. The conclusion from the study of Abrahams and Parsons (1977) was that these soils could supply a significant proportion of the recommended intake (UK Dept. of Health) for some minerals nutrients, up to 17% in the case of Fe.

A further study by Abrahams (1997) on 13 geophagic soils from Uganda, many of which were sold in local market as medicines showed that they had a high (>50%) clay mineral content dominated by kaolin minerals (85%).

Geissler et al. (1999) who studied geophagy among pregnant Ugandan women reported that most women (73%) were eating soil at a median intake of 41.5 g (range 2.5-290 g) per day and that most women (84%) ate soil at least once a day. When asked why they eat soil, 26 women (68%) answered that they "like it" or that it is tasty and that eating it "felt nice" in the mouth and satisfied them.

Geissler et al. (1997) in a previous study on geophagy among school children in western Kenya Nyanza province reported that over 70% of a sample of school children consumed soil at an average rate of ~30 g daily. The sources of the geophagy materials were from termite mounds, edges of paths and gullies, weathered stones and walls of huts.

The chemistry and mineralogy of geophagic soils from Zimbabwe, North Carolina- USA, and Hunan Province- China, were studied by Aufreiter et al. (1997). The soils from Zimbabwe which came from termite mounds contained kaolinite as the main clay mineral. American samples were dominated by halloysite while the Chinese samples had an abundance of smectite. These workers concluded that the results were consistent with the effects desired by the consumers of the soils, namely alleviation of diarrhoea (Zimbabwe), unspecified health benefits (North Carolina) and as famine food (China).

Geophagy among Animals

Geophagy is quite common among animals. The exact reasons why animals consume soil are not clear even though many hypothesis pertaining to the beneficial aspects of soil ingestion have been advanced. Wilson (2003) has summarized some of the more recent work carried out on geophagy in the animal kingdom.

As in the case of human geophagy, clay mineralogy is the more dominant factor in animal geophagy as well. Two studies on geophagy in birds (Diamond et al., 1999; Gilardi et al., 1999) concluded that from among the clay minerals consumed, smectite was dominant. The geophagic materials were assumed to reduce the bioavailability of poisonous or bitter-tasting compounds such as alkaloids and tannic acids in fruits and seeds.

Cattle and elephants (Figure 10.2) are among the large mammals studied for their geophagic habits (Klaus et al., 1998; Abrahams, 1999; Mahaney et al., 1996). In these cases, there was no definite conclusion as to the geophagy, but prevention of gastrointestinal upsets, alleviation of sodium deficiency and acidosis were speculated.

Fig. 10.2. Elephants feeding at a soil lick site in the Udawalawe National Park, Sri Lanka (photo courtesy of Enoka Kudavidanage)

Geophagy in horses has also been regarded as a sign of nutritional deficiency (Ralston, 1986). McGreevy et al. (2001) carried out a geochemical analysis of 13 equine geophagic sites from different parts of Australia. They observed higher concentrations of iron and copper in the soils eaten as compared to the control sites and suggested that these elements provide the stimulus for geophagia. What was interesting also, was how the geophagic horses may have sensed the increased concentrations of these elements in soils.

Gorillas, monkeys and chimpanzees have been the subject of several investigations for their geophagic habits. Subsamples of termite mound soil used by chimpanzees for geophagy and top soil (never ingested by them) from the forest floor in the Mahale Mountains, National Park, Tanzania were analyzed (Mahaney et al., 1999) to determine the possible stimulus or stimuli for geophagy. Here too, the ingested samples had a dominant clay texture, equivalent to a clay stone. Interestingly the clay mineral metahalloysite found in these clays eaten by the chimpanzees, is also an ingredient in the pharmaceutical Kaopectate TM, used to treat minor gastric ailments in humans. The fact that these soils had a high pH range of 7.2 to 8.6 also indicates a possible antacid property of the eaten soils.

Ketch et al. (2001) argue that these chimpanzees selectively consume certain plant species for their medicinal value, and that they could also perhaps engage in geophagy for similar reasons.

Several case studies of geophagy among several species of monkeys and gorillas (Mahaney et al., 1993; Mahaney et al., 1995; Bolton et al., 1998; Heymann and Hartmann, 1991; Setz et al., 1999; Mahaney, 1993) indicate a clay mineral dominance and a possible ability to absorb toxins and function as a mineral supplement.

Wilson (2003) critically reviewed the studies mentioned above. He was of the view that there is no single explanation for the geophagic behaviour in animals, including man. What was lacking in these studies was the insufficient characterization of the mineralogy and chemistry of the geophagic materials. He suggested that the following information is required for such geochemical characterization:

(a)   The nature of the constituent kaolin minerals and their abilities to coagulate, aggregate and/or dispense under acid conditions.

(b)   The occurrence of interstratified kaolinite-smectite and whether this mineral could account for the anomalously high cation exchange capacities (CEC) of geophagic soils/clays.

(c)   The nature of the smectite minerals, particularly the occurrence and extent of interstratification with other layer silicates or interlayering with non-exchangeable Al-hydroxides, since this relates to the ability of the clay to adsorb potential toxins.

(d)   Characterization of the iron oxide minerals and determination of the ratio of reactive to non-reactive species in relation to supplementation of Fe in the diet.

(e)   Occurrence and nature of aluminous minerals and whether they function effectively as natural antacids.

It appears likely that the pleasant touch of the ingested clay is the initial stimulus for geophagy in animals and that it subsequently becomes a learned behaviour. It could also well be that clays have many useful medicinal properties which the animals use for a more general and all round well being rather than for a specific ailment.

INGESTION OF GEOMATERIALS FOR HUMAN HEALTH- THE MEDICAL CONCERNS

In spite of the seemingly beneficial aspects of geophagy as discussed above, there is undoubtedly much apprehension and debate about the benefits and the toxicity of ingestion of geomaterials. Tateo et al. (2001) attempted to study the laboratory simulation of the digestive process using 14 different herbalists' clays for internal use as found in the Italian market. The digestion experiment consisted of two stages (i) involving acidic solutions (matching the stomach environment) (ii) reproducing bile-pancreas juices. These authors observed that one effect of clay digestion is the dissolution of carbonates, transferring Ca and Mg from the solid to the digestive solutions along with an increase of pH.

As shown in Figure 10.3 there is a bimodal distribution of pH, from 1 to 3 and from 5 to 6. Some elements such as Al, P, Be, Sc, V, Fe, Cu, Zn, Ga, Ba, La, Nb, and REE were associated with a lower pH and others such as Ca, Sr, Mg with higher pH. The element concentrations in the final solution were then compared with the maximum daily dose as given by drinking water regulations. Among the elements not hosted into the carbonates, Al showed high concentrations. Tateo et al. (2001), on account of the hazardous chemical elements that were detected, warned of the risk involved in using geomaterials for health and suggested that further multidisciplinary research be conducted with clay mineralogists, toxicologists and physiologists.

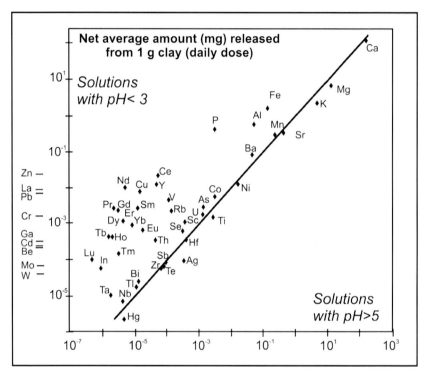

Fig. 10.3. Average concentration of analysed elements released into solution from 1 g of clay at different pH with clay/water at 1:10 (Tateo et al., 2001)

PICA (Persistent Ingestion of Non-Nutritive Substances) of clay or geophagy is known to bind potassium in the intestine leading to severe hypokalemic myopathy (Ukaonu et al., 2003). It may also cause a number of serious conditions such as heavy metal poisoning, hyperkalemia, iron and zinc deficiency, vitamin deficiency, intestinal obstruction or perforation, dental injuries and parasitic infestations (Rose et al., 2000; Federman et al., 1997). Pregnant women who eat 1 lb of clay per week have been known (Simulian et al., 1995). Woywodt and Kiss (1999) reported a case of an adult, non-pregnant African woman with perforation of the Sigmoid colon due to geophagia. Since geophagic patients do not normally wish to divulge this clay-eating habit, plain X-ray films are the diagnostic procedure for soil detection in the human body (Fig 10.4).

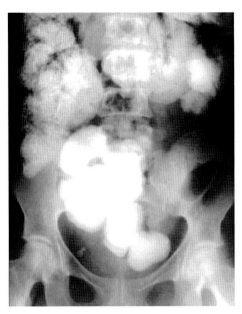

Fig. 10.4. Plain abdominal X-ray film of the patient on admission to the hospital. Free intra-abdominal air is present and the entire colon appears to be densely impacted with radio opaque material that proved to be dry soil (Woywodt and Kiss, 1999; copyright ©1999, American Medical Association. All Rights Reserved)

PODOCONOSIS-A GEOCHEMICAL DISEASE

Podoconiosis is a disease characterized by swelling and deformity of the legs associated with enlargement of the draining lymph nodes (Figure 10.5). It is also termed non-filarial elephantitis (Price, 1988, 1990). This disease is predominantly found in regions of tropical Africa where fine reddish brown soils are prevalent and in areas of high altitude (>1250 m), modest average temperature (20°C) and high hot season rainfall (>1000 mm annually) (Price and Bailey, 1984). This epidemiological relationship is seen in the Wollamu district of Ethiopia, Nyamkere range of Kenya, Tanzania, Rwanda, Burundi, Cape Verde Islands and Cameroon Highlands (Harvey et al., 1996). The special geochemical features as seen in these tropical regions clearly imply a geographical specificity to Podoconiosis.

The superimposition of the prevalence data for the disease showed a correlation where the areas coincide with a distribution of alkali basalt rocks. The extremely fine particles (<5 μm) seen at these sites are formed by the intense rock weathering so typically seen in the tropical lands.

Fig. 10.5. A young Ethiopian with bilateral idiopathic elephantiasis and marked lymphostatic verrucosis on both feet and a view of the foot of an African from Zimbabwe. There is thick diffuse hyperkeratotic swelling, again confined to the lowest part of the leg (photo by Dr. E.W. Price, Copyright Springer, Germany).

Interestingly, the analyses of particles from a digest or a diseased femoral lymph node (Price and Plant, 1990) showed an identical size range (i.e. 0.3-6 μm) which suggested that either the toxic particles are within this size range or that a smaller size facilitated the uptake into the tissues. The analysis of these microparticles showed that they consisted mainly of Si and Al, with lesser amounts of Ti and Fe suggesting the possible presence of the aluminum silicate kaolinite, silica in the amorphous state with some quartz and iron-oxide. The natives of these areas which have very fine soil mineral particles walk barefooted and it has been suggested that the abrasions in the feet caused thereby serve as entry points for the tiny particles, which finally enter the lymph nodes (Price and Henderson, 1978; Blundell, et al., 1989).

As discussed in the review of the geochemical factors linked to podoconiosis by Harvey et al. (1996), Hochella (1993) noted that the toxicity of a mineral is related to:

(a) Mechanical or dimensional properties, namely the size of particles in diseased tissues. Smaller particles are rapidly absorbed through the abrasions of feet and transported in lymphatics. In the case of other particles, a specific fibrous shape is the most important toxicological determinant (Stanton et al., 1981). No preponderance of fibrous minerals was observed in the case of podoconiosis.

(b) With decreasing size, the surface area to volume ratio of particles increases and different chemical properties gain importance. Among these factors which determine the ability of particles to interact in biological systems are (i) surface and near surface composition, (ii) surface atomic structure, (iii) surface microtopography, (iv) surface charge, (v) pH, (vi) dependence on surrounding solubility, (vii) durability (dissolution rate) and (viii) associated minor and trace elements.

The nature of the interaction of these mineral phases with the effector cells of the immune system, notably the macrophage, finally determines the ability of the mineral to produce the diseases. This is termed the pathogenicity of the mineral. The cell takes up the mineral particles which then are activated to produce many chemical species which in turn damage the biomolecules. Finally, inflammation and fibrosis result. If the cells die as a result of the activation mentioned earlier, the mineral particles are then available for re-uptake and the process continues (Driscoll and Maurer, 1991; Gabor et al., 1975; Kennedy et al., 1989).

A disease, termed Kaposi's Sarcoma (KS) is very similar to podoconiosis and is also found mostly in Africa. Ziegler (1993) noted that the prevalence of both podoconiosis and KS in highland areas of Africa, close to volcanoes is suggestive of a shared pathogenetic relationship due to exposure to volcanic soils. KS is believed to arise in the lymphatic endothelium and is associated with chronic lymphoedema. Figure 10.6 illustrates the geographic distribution of Kaposi's Sarcoma in Africa and relation of ultra basic basalt provinces and prevalence of podoconiosis.

Podoconiosis and KS are excellent examples of diseases which clearly illustrate the interrelationship between Geology and Health in tropical areas where the pathways of chemical elements and minerals from the immediate environment to the body is direct. The term "geochemical diseases" is attributable to such cases where the aetiology of a certain disease is almost directly and clearly related to a characteristic elemental composition of the immediate geological environment.

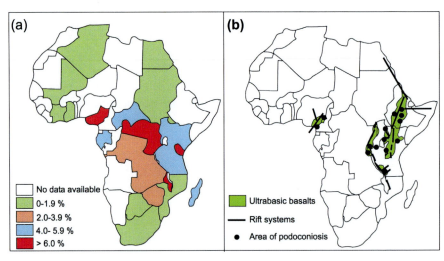

Fig. 10.6. (a) The geographic distribution of Kaposi's Sarcoma in Africa. Proportional rates (percent KS divided by total cancer over the reporting period) in African males prior to 1981; (b) Relationship of ultra basic basalt provinces and prevalence of podoconiosis in the East African Rift System (Ziegler, 1993)

NATURAL DUST AND PNEUMOCONIOSIS

Direct inhalation of very fine soil particles (2 to 63 μm) is another route by which minerals may get into the human body. Derbyshire (2003) reported a case study in High Asia (Himalayas-Karakoram and Tian Shan-Pamir areas) where whole communities of people were exposed to the adverse effects of natural (aerosolic dust). The exposure levels had reached those normally found in areas of high-risk industries.

The reasons for this area being the world's most efficient producer of silty debris were:
- (a) high known uplift rates (up to 12 mm/yr)
- (b) unstable slopes
- (c) glaciations and widespread rock break up during freezing
- (d) soil weathering due to hydration.

Further, the characteristic properties of these loess included high porosity and collapsibility on wetting. Mineralogically, they consisted of angular, blade-shaped quartz grains (~60-65%) with minor feldspars, micas and carbonates. Clay minerals were in the range of 12-15%.

The prevalence of silicosis, as studied by the use of chest X-rays, showed 1.03-10% cases in the Gansu Province, China. In some parts of India, too, inhalation of natural dust has resulted in a high prevalence of silicosis. Norboo et al. (1991) and Saiyed et al. (1991) reported cases of progressive massive fibrosis. In the Ladakh region in India, where these studies were done, wind-blown loessic dust, including aggregates of the finest fractions was a clear health hazard. Adults in the age range 50-62, had up to 45% silicosis problems. Bulk chemical analysis of lung tissues had showed that over 54% of the inorganic dust was silica.

The case studies described in this chapter illustrate the fact that human beings and animals are only a part of the large scale geochemical cycles of elements. The geochemical pathways of the physical environmental link up with the biochemical and physiological pathways in the human body. Medical geology is concerned with total geochemical cycling and further information is now emerging on the roles played by minerals within the human body.

CHAPTER 11

HIGH NATURAL RADIOACTIVITY IN SOME TROPICAL LANDS – BOON OR BANE?

Natural radiation has been in existence from the time planet earth started to form and there are about 60 radionuclides present in nature. These are found naturally in air, water, soil, rocks, minerals and food. About 82% of this environmental radiation is from natural sources, the most important of which is radon.

The United Nations Scientific Committee on Effects of Atomic Radiation Report (UNSCEAR, 2000) states that while the natural background radiation is the largest contributor to human exposure, the worldwide average annual effective dose per capita is 2.4 mSv. Some radionuclides such as ^{238}U, ^{232}Th and ^{40}K among others are important as naturally occurring radionuclides and geological and geochemical processes play the major role in their distribution in nature. These are found in some minerals such as monazites and zircons. Some areas of the world have anomalously high background radiation and these are called high background areas (HBRAs) (Table 11.1). In these areas, the geology and geochemistry of the source rocks and minerals have the greatest influence on the localization high natural radiation.

In radioactive material measurement, the International System of Units (SI) is generally used. However, there are several other units that are used in the literature and this causes a certain amount of confusion among the readers. The common reference measurements are radioactivity and radiation dose. These units are often referred to in both their SI units (Becquerel and sievert) and the traditional units (curie and rem). The University of Ottawa (2007) has created the conversion between the two systems as given in http://www.uottawa.ca/services/ehss/ionizconversion.htm.

Table 11.2 shows the mineral sources of the naturally occurring radioactive materials. Extreme HBRAs are found in Guarapari (Brazil), Southwest France, Ramsar (Iran), Kerala coast (India) and Yangjiang (China) (Table

11.1). Of these the majority of HBRAs are found in tropical, arid and semiarid areas.

Table 11.1. Natural Radioactivity in some selected areas (based on the UNSCEAR Report, 1993)

Area	Mean (mGy/year)	Maximum (mGy/year)
Ramsar, Iran	10.21	(260)
Guarapari, Brazil	5.52	(35)
Kerala, India	3.82	(35)
Yangjiang, China	3.51	(5.4)
Hong Kong, China	0.67	(1.00)
Norway	0.63	(10.5)
France	0.60	(2.20)
China	0.54	(3.01)
Italy	0.50	(4.38)
World average	0.50	
India	0.48	(9.6)
Germany	0.48	(3.8)
Japan	0.43	(1.26)
USA	0.40	(0.88)
Austria	0.37	(1.34)
Ireland	0.36	(1.58)
Denmark	0.33	(0.45)

[1]High Levels of Natural Radiation 1996, M. Sohrabi, p 57-68 Elsevier Science B.V, 1997)
[2]1982 UNSCEAR report

Terrestrial Radiation in Beach Sands in Brazil

In certain beaches in Brazil, monazite sand deposits are abundant. The external radiation levels on these black beach sands range up to 5 mrad/hr (50µ Gy/hr) and is nearly 400 times the normal background in USA. The Brazilian coastal sands have several minerals among which are monazite, rutile, ilmenite, zircon, cassiterite, thorianite, pyrochlore and niobate- tantalite. Ilmenite ($FeTiO_3$)-96%, rutile (TiO_2) - 0.5%, zircon ($ZrSiO_4$)-2.5% and monazite- $REE_3(PO_4)$ rare earth phosphate- 3% comprise the main minerals. The monazite from the beach sands of Brazil contain up to 39% cerium oxide (CeO_2), 16% of lanthanum oxide (La_2O_3), 14% of neodymium oxides (Nd_2O_3), 5% yttrium, 6% thorium dioxide (ThO_2) and 0.31% uranium oxide and phosphates.

Fujinami et al. (1999) studied the exposure rates from terrestrial radiation at Guarapari and Meaipe in Brazil and made measurements of absorbed dose rate in air one meter above the ground and on the surface with a portable NaI (Tl) scintillation survey meter. The highest dose rate was 0.6 µGy/h at one meter height in the streets and the highest rate at one meter height on the beach was 6.2 µGy/h and on the surface of the beach it was 15 µGy/h. Interestingly the beach at "Areia Preta"-meaning black sand, is highly sought after by Brazilian people for their alleged health benefits (Eisenbud, 1973). In an area on the beach at "Areia Preta" where black sand was not found, the absorbed dose rate was 0.04 µGy/h and in the Meio Beach where white sands are found and separated from the Areita Preta beach, the dose rate was 0.09 µGy/h.

Table 11.2. Ores containing naturally occurring radioactive materials (Hipkin and Shaw, 1999)

Ore	Main Constituent	Typical Active Concentrations (KBq/kg)	
		Thorium-232	Uranium-238
Phosphate rock	Calcium phosphate	0.1	1.5
Ilmenite	Iron titanium oxide	1	2
Rutile	Titanium dioxide	0.2	0.2
Rare earth concentrate	Cerium oxide	5	0.1
Baddeleyite	Zirconium oxide	1	10
Zircon	Zirconium silicate	0.6	3
Pyrochlore	Niobium oxide	80	10
Monazite	Cerium phosphate	300	40

At Meaipe beach, the highest dose rate in the street area was 5.2 µGy/h while on the surface of the beach near the breakwater the highest value reached a level of 32 µGy/h.

Lauria and Godoy (2002), reported anomalously high ^{228}Ra concentrations up to 2 Bq/L in waters of a coastal lagoon close to a monazite sand separation plant in Buena lagoon area in the Rio de Janeiro state in Brazil. Even though it had earlier been suggested that this was caused by mineral processing, it was later concluded that the abnormal radium concentrations had a natural origin from springs at the lagoon head with high ^{228}Ra and ^{226}Ra concentrations.

Strong relationship among radium and light rare earth elements (LREEs), ^{228}Ra/^{226}Ra activity ratio and the REE pattern were suggestive of monazites as the main source of nuclides in the water. It was suggested that low pH

(3.7) and high salinity (14‰) caused a disturbance of the chemical stability of monazites. There was a relatively low mobility of thorium but a high mobility of radium and LREEs.

In other parts of the world, the radium concentration in surface waters normally range from 0.01 to 0.1 Bq/L (Iyengar, 1990). In the Buena lagoon area in Brazil, the ^{228}Ra concentration in the spring waters varied between, 1.7 and 2.5 Bq/L and for ^{226}Ra it was 0.5 Bq/L. The LREE concentrations in the water were also very high with La = 50 µg/L, Ce = 110 µg/L and Nd = 60 µg/L.

The recent study by Dias da Cunha et al. (2004) on the airborne particles in the Buena village using lichen samples showed that during the last 15 years, the inhabitants of the village had been exposed to monazite particles. The result form aerosols and lichens also suggested that the swampy area is a source of ^{226}Ra and ^{210}Pb-bearing particles in addition to the monazite dust.

In Morro do Ferro in the Pocos de Caldas Plateau of Brazil is an uninhabited hill 0.35 km^2 in an area consisting of large rare earth ore deposits that rise to above 250 meters from the surrounding area. The radiation levels here are 1 to 2 mrad/hr (0.01 to 0.02 mGy/hr) over the 0.35 km^2 area and are between 100 to 200 times the non- elevated natural gamma radiation background (Paschoa and Godoy, 2002).

Monazite Rich Beach Sands of India

As shown in Table 11.3 some of the beach deposits of India, notably the Kerala coast Tamil Nadu coast, Bhimilipatanam (Andhra Pradesh) coast and Chhatrapur (Orissa) coast are prominent HBRAs. As in the case of Brazil, the high levels of natural radiation is due to the abundance of monazite along with other heavy minerals such as ilmenite, rutile, zircon and garnet among others. The coastal regions of peninsular India have therefore received great international attention on account of the high natural radioactivity (Paul et al., 1998).

Along the 570 km long coastline of Kerala in the Southwest of India, there are major deposits of monazite-rich heavy minerals with very high natural radiation. The monazite deposits are larger than those in Brazil and the dose from external radiation, is on the average similar to doses reported in

Brazil, 5-6 mGy/yr. Individual doses up to 32.6 mGy/yr, however, have also been reported.

Table 11.3. Comparison of radiation dose rates of Chhatrapur beach placer deposit area and other high background radiation areas (HBRAs) (after Mohanty et al., 2004a)

Area (Country)	Characteristics of area	Absorbed dose rate in air (nGy/h)
Guarapari (Brazil)	Monazite sands	90-90,000
Morro Do Forro (Brazil)	Th-rich alkaline intrusives	2800
Yangjiang (China)	Monazite sands	370
Nile Delta (Egypt)	Monazite sands	20-400
Southwest (France)	Uranium minerals	10-10,000
Ramsar (Iran)	226Ra in spring water	70-17,000
Cox's Bazar (Bangladesh)	Monazite sands	260-400
India		
Kerala coast	Monazite sands	200-4000
Ullal (Karnataka)	Monazite sands	2100
Tamil Nadu coast	Monazite sands	200-4000
Kudiraimozhi (Tamil Nadu)	Monazite sands	200-900
Bhimilipatanam (Andhra Pradesh)	Monazite sands	200-3000
Kalpakkam (Tamil Nadu)	Monazite sands	3500
Ayirmamthengu (Kerala)	Monazite sands	200-1400
Neendakara (Kerala)	Monazite sands	200-3000
Chhatrapur (Orissa)	Monazite sands	375-5000

Paul et al. (1998) studied 4 different locations (Fig 11.1) in coastal India for their natural radiation and the results are shown in Table 11.4. The radiation and fields ranged from 200 to 3000 nGy/h at the different sites, with large radiations in the fields, in some cases up to an order of magnitude. The Neendakara and Bhimilipatnam deposits of the Kerala coast had the highest levels and were associated with the beach placer deposits. It was also observed that with the onset of tropical monsoon, higher deposition of heavy minerals takes place and hence the enhanced natural radiation at these beaches.

Mohanty et al. (2004a, 2004b) carried out some studies on the natural radioactivity of high background radiation areas on the eastern coast of

Orissa, India. They found that the average activity concentrations of radioactive elements such as ^{232}Th, ^{238}U, and ^{40}K were 2825±50; 350±20 and 180±25 Bq/kg, respectively for the bulk sand samples. The absorbed gamma dose rates in air due to the naturally occurring radionuclides varied from 650 to 3150 nGy/h with a mean value of 1925±718 nGy/h This is very much higher than the world average value of 55 nGy/h (UNSCEAR, 1988). The presence of ^{232}Th contributed a maximum of 91% (1750 nGy/h) to total absorbed dose rate in air followed by ^{238}U of 8.5% (165 nGy/h) and by ^{40}K of 0.5% (8 nGy/h). The annual external effective dose rates for the region varied from 0.78 to 3.86 mSv/yr with a mean value of 2.36±0.88 mSv/yr.

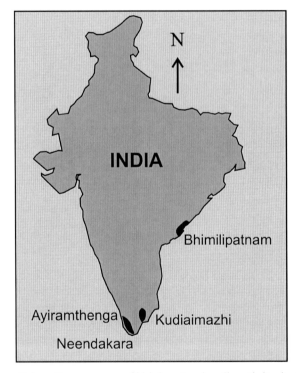

Fig. 11.1. Locations of some areas of high natural radioactivity in India (Paul et al., 1998)

This area is therefore a high background radiation area similar to other radiation areas of southern and southwestern coastal regions of India (Mohanty et al., 2004b). Similar results were obtained by Mohanty et al. (2004a) for the Chhatrapur beach placer deposit, also of Orissa, India. Here too, the enhanced levels of radiation were due to monazite and zircon sands.

Table 11.4. External gamma radiation fields (Mohanty et al., 2004a)

Site	No. of Locations Monitored	Radiation Field Range	(nGyh^{-1}) Mean±SD
Ayiramthengu	64	200-1400	300±270
Neendakara	64	200-3000	500±330
Kudiraimozhi	43	200-900	370±140
Bhimilipatnam	40	200-3000	430±330

High Natural Radioactivity of the Minjingu Phosphate Mine, Tanzania

The Minjingu phosphate mine in northern Tanzania located to the east of the saline lake Manyura at Arusha is a HBRA. Phosphate deposits are known to contain significant traces of the long-lived radioelements ^{238}U, ^{232}Th and ^{40}K. These contribute to the internal and external radiations to man (Halbert et al., 1990; Bettencourt et al., 1990). There are about 2000 inhabitants in the village of Minjingu and these people may be exposed to naturally occurring radionuclides from the emissions caused by open dry phosphate mining (Banzi et al., 2000). Bianconi (1987) had earlier reported that the deposit has uranium activity with a maximum concentration of 800 mg/kg U_3O_8. It had been observed that the average dose rate in air of 1415 nGy/h is 28 times the average terrestrial dose rate of 50 nGy/h in air worldwide (UNSCEAR, 1982). Further, the estimated average effective dose over 5 years at Minjingu is about 12 mSv/y which is 12 times the allowed average dose limit of 1 mSv/y for members of the public (IAEA, 1996). Based on these observations the Minjingu phosphate mine is rated as an area ranked with the highest natural radiation background recorded in the world (UNSCEAR, 1982).

Very High Natural Radiation in Ramsar, Iran

UNSCEAR (2000) has classified Ramsar, a city located in northern Iran, as an area with one of the highest natural radiation levels in the world. The city lies between the Elburz Mountains and the Caspian Sea at an average elevation near sea level, and is a popular tourist resort. The natural radiation in some locations at Ramsar is 55-200 times higher than the normal background areas. The annual exposure levels in areas with elevated levels of natural radiation in Ramsar reach values as high as 260 mGy/y and average exposure rates are about 10 mGy/y for a population of about 2000 residents (Mortazavi et al., 2001).

The origin of the high levels of natural radioactivity is shown in Figure 11.2. The radioactivity in the high level natural radiation areas of Ramsar is due to ^{226}Ra and its decay products brought up to the earth's surface by the water of hot springs, of which there are at least 9 with different concentrations of radium.

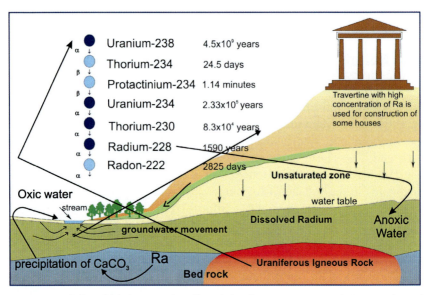

Fig. 11.2. Origin of high natural radioactivity in Ramsar, Iran (Mortazavi et al., 2001)

When the groundwater reaches the surface at hot spring locations, travertine, a calcium carbonate ($CaCO_3$) mineral precipitates out of solution with dissolved radium substituting for calcium in the mineral as $RaCO_3$. High concentrations of radium carbonate are found in the residual deposits (up to 30 m thick) of the hot springs. These materials are often used as building materials by the residents of the area. The gamma radiation at waist level in these houses reaches up to about 20 μSv/h. The walls of the bed rooms (made of sedimentary rocks from the hot springs) of these dwellings also show dose rates as high as 143 μGy/h. In spite of the fact that the levels of natural radiation in these areas is up to 200 times higher than normal background levels, nearly all inhabitants still live in their old ancestral dwellings, without any ill effects (Mortazavi, 2003).

High Natural Background Radiation in Yangjiang, China

The Yangjiang area in China is also classified as a HBRA and has dose rates about 300-400 mrad/yr (3-4 mGy/yr). As in the case of Kerala and Orissa in India and Guarapari in Brazil, the main source of the radiation is monazite containing Th, U, and Ra.

The Oklo Natural Reactor

A most interesting geological phenomenon occurred during the Proterozoic period (around 1800 my) when a natural nuclear reactor was formed at Oklo in Gabon, West Africa. This was discovered quite by accident, when in 1972 some French scientists working on a shipment of uranium from Oklo observed that it had a much smaller proportion (as low as 0.440%) of ^{235}U than normal (Cowan, 1976). Natural uranium always has the identical isotopic composition ^{238}U (99.27%), ^{235}U (0.72%) and traces of ^{234}U. Of these, only ^{235}U can take part in chain reactions and it is therefore very closely studied.

Since the ^{235}U isotope is present only in low abundance (0.72%), enrichment up to 3% or greater is required for use in commercial nuclear reactors. During the Proterozoic period around 2000 my ago however, this critical level of 3% was present in nature and this was sufficient for a sustained nuclear reaction. The abundant groundwater present at the Oklo site served as a moderator, reflector and cooler for the fission reaction. The presence of neodymium provided strong evidence that the natural reactor had indeed operated and that it had consumed six tons of uranium over several hundred thousand years releasing 15,000 MW years of energy.

Lovelock (1988), described the geological process that acted as a natural reactor (Figure 11.3). Uranium is known to be readily soluble in the presence of oxygen and during the Proterozoic times, when oxygen was present, the groundwater was oxidizing. Under these conditions, uranium dissolved yielding the uranyl ion which was present in trace quantities. At the Oklo site, a stream had flowed into an algal mat which had the microorganisms having the capability of concentrating uranium specifically. The efficiency of this process of uranium concentration by the microorganisms was so high that uranium oxide was deposited in the pure state enabling a nuclear reaction to start. The water present enables the slowing and reflecting of neutrons so that it is controlled and self-regulating (Lovelock, 1988).

The unique feature about the Oklo natural reactor is that it was functional as far back as 2000 my ago and that it was a complete natural geochemical and microbiological phenomenon that was entirely self regulating. It was after a period of nearly 2 billion years that, modern scientists were able to duplicate nature's natural reactor at Oklo and produce the first man-made nuclear reactor.

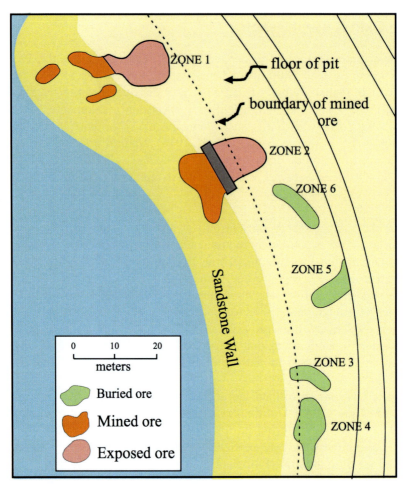

Fig. 11.3. Disposition of six of the Oklo, 2 billion year old natural reactors (from Anomalies in Geology, Science Frontiers Jan-Feb 1999 (Corliss, 1999)

RADIATION AND HEALTH

Radiation is a form of energy and which comes from man-made sources such as X-ray machines or from natural sources such as the sun, outer space and radioactive minerals in the environment. Of the naturally occurring elements, some, known as radionuclides emit energy in the form of alpha rays, beta rays and gamma rays. These rays interact with surrounding matter which results in ionization. When this type of ionization radiation enters the human body, it is referred to as internal exposure, as opposed to external exposure from radiation outside the body, such as from the sun or from radioactive materials. When the radiation is absorbed in the human body, the resulting chemical reactions can cause adverse health effects which may be observed in some cases after many years. These adverse health effects take the form of skin diseases, cancer and even death. About 80% of human exposure is from natural sources while the balance is from man-made sources (Table 11.5).

It is therefore the natural background which has evoked a great deal of interest in environmental health. Figure 11.4 illustrates the importance of the natural background radiation as against anthropogenic causes. The radiation encountered from natural radiation is several orders of magnitude compared to those emitted by nuclear fall outs, and medical diagnostics. The intriguing question therefore is "what is the effect of natural background radiation on human health?"

It is interesting to note that in all the case studies such as Guarapari (Brazil), Ramsar (Iran), Yangjiang (China), Orissa and Kerala (India), the adverse health effects appear to be very low. On the contrary, in some cases the population exposed to these HBRAs appears even healthier with higher life spans as compared to the control areas.

An epidemiological survey carried out in the Yangjiang HBRA in China has revealed that there are no observed adverse health effects based on cancer mortality data from 1,008,769 person-years in HBRA and 995,070 person-years in control areas (Zha et al., 1996). The study also included hereditary diseases and congenital malformations, human chromosome aberrations and immune functions.

Between 1970 and 1986 the Yangjiang HBRA population (background radiation 5.5 mSv/y) was compared with people living in two adjacent low-background counties Enping and Taishan (background radiation 2.1

mSv/y), the cohorts being 74,000 (Yangjiang) and 77,000 (Enping and Taishan). For the age group of 10 to 79, the results showed that the general cancer mortality was 14.6% less in Yangjiang compared with Enping and Taishan. The leukemia mortality was 16% lower in men and 60% lower in women from Yangjiang.

Table 11.5. Man's exposure to ionizing radiation (source: Australian Radiation Protection and Nuclear Safety Agency, 2004)

Source of Exposure	Exposure
Natural Radiation (Terrestrial and Airborne)	1.2 mSv per year
Natural Radiation (Cosmic radiation at sea level)	0.3 mSv per year
Total Natural Radiation	1.5 mSv per year
Seven Hour Aeroplane Flight	0.05 mSv
Chest X-Ray	0.04 mSv
Nuclear Fallout (From atmospheric tests in 50's & 60's)	0.02 mSv per Year
Chernobyl (People living in Control Zones near Chernobyl)	10 mSv per year
Cosmic Radiation Exposure of Domestic Airline Pilot	2 mSv per year

In Ramsar, Iran where the natural radiation, as explained earlier was very high, some epidemiological studies (Mortazavi, 2003), have shown no significant adverse health effects. Preliminary results on chromosome aberrations showed no significant difference among those who lived in houses with extraordinarily elevated levels of natural radiation. A cytogenetical study was carried out on 21 healthy inhabitants of the high level maturation radiation areas and 14 residents of a nearby control area. There was no observed positive correlation between the frequency of chromosome aberrations and cumulative dose of the inhabitants. In the case of hematological parameters too, there was no statistically significant alteration in the very high level natural radiation areas (VHLNRAs) of Ramsar as compared to those living in a neighbouring control area.

High doses of ionizing radiation are known to suppress the activity of the immune system, whereas low-level whole body irradiation (WBI) can enhance the immunological response. This was tested in Ramsar (Mortazavi, 2003) and the preliminary results indicated that relatively high doses are not immunosuppressive. The results however, were not conclusive.

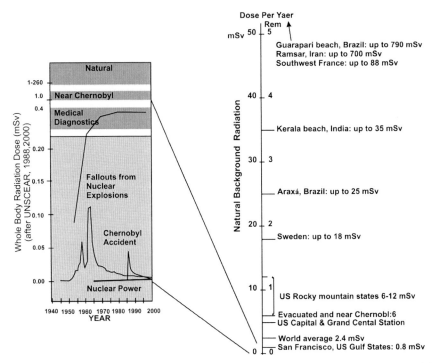

Fig. 11.4. Radiation dose received from natural and human sources (Jaworowski, 2000)

Adaptation to high levels of natural radiation has been the subject of many studies (Olivieri et al., 1984; Sanderson and Morely, 1986; Kelsey et al., 1991; Rigand et al., 1993; Joiner 1994; Azzam et al., 1994). In order to assess the possible induction of adaptive response in the inhabitants of HLNRAS of Ramsar, blood samples of the residents and a nearby control area were exposed to a challenge dose, of 1.5 Gy (natural radiation was used as the adapting dose). The results showed 56% fewer MCAPC (mean number of chromosomal aberrations per cell) caused by the challenge dose among HLNRA s as compared to lymphocytes from residents of low background areas (Ghiassi-nejad et al., 2002). It had been reported that the Ramsar results suggest that chronic low dose radiation may not only reduce mortality form all causes and cancer mortality, but may also be protective against accidental high dose radiation (Pollycove and Feinendegen, 2001).

In Kerala India, another area of high background natural radiation, it has been reported that over 140,000 inhabitants receive an annual average dose of 15-25 mbq (Kesavan, 1996). The average life span of the inhabitants of

Kerala is 72 years while for all India it is only 54 years (Goraczko, 2000). It has been shown by Nair et al. (1999), following a detailed study, that the residents of HLNRA s of Kerala showed no evidence of enhanced rates of cancer. Further, there was no correlation with the natural radiation levels in the densely populated areas of the monazite-rich Kerala sands with the stratification of newborns with malformations, still births or twinning (Jaikrishan et al., 1999).

The observation that there are no noteworthy adverse health effects in some of the worlds very high level natural radiation areas brings into question the 'safety levels of radiation'. Jaworowski (2000) in his presentation on "Ionization radiation and radioactivity in the 20th century" summed up his views as follows.

"Man's contribution to the contents and flows of radionuclides and of radiation energy in the particular compartments of the environment consists of a tiny fraction of the natural contribution. In some areas in the world, the natural radiation doses to man and to other biota are many hundreds of times higher than the currently accepted dose limit for the general population. No adverse health effects were found in humans, animals and plants in these areas. In the future reconstruction of the edifice of radiation protection that now stands on the abstract LNT foundations, down-to-earth approach will be necessary taking into account apparently safe chronic doses in the high natural radiation areas, rather than the statistical variation around an average global value. It seems, therefore, that studies of these areas deserve special attention and support in the coming years".

Man still has to learn a lot from nature –
Geology provides the clues!

CHAPTER 12

BASELINE GEOCHEMICAL DATA FOR MEDICAL GEOLOGY IN TROPICAL ENVIRONMENTS

Baseline geochemical data and geochemical maps are now proving to be extremely useful tools in epidemiology and medical geology. Information on the geochemical pathways and regional distribution of essential and toxic elements in the environment is now more easily available and there is greater understanding of the impact of trace elements in health and disease, their bioavailability and homeostasis. With the help of statistically significant data, it is now possible to pin point areas vulnerable to diseases that are clearly geographically and geologically linked.

Geochemical maps showing detailed distribution of various chemical elements in soils, stream sediments and water are now available in most developed countries. In developing countries however, due to a variety of factors such as lack of high precision instrumentation, funding, trained personnel, rugged and inaccessible terrains, among others, such maps are still not available. China, however, has shown remarkable progress in geochemical mapping since 1979, and there has been more progress in China than anywhere else in the world. It seems highly probable that from 1980 onwards, a larger area has been sampled in China than in all other countries combined (Darnley, 1995; Xie and Cheng, 2001).

Even though these geochemical maps were primarily used for mineral exploration, they are now used extensively as in the case of China, Finland, Sweden and Norway, for geomedical purposes (Tanskanen et al., 1988; Selinus, 1988). The delineation of the high-risk areas based on the levels of abundance of essential and toxic elements is done with this baseline data and when combined with geological information such as lithology, mineralogy and geomorphology, *'geochemical maps'* could be prepared. Many studies on medical geochemistry are carried out using these geochemical maps (Dissanayake and Chandrajith, 1999).

In view of the fact that clear associations and correlations between the geological environment and human and animal health is far more visible in developing countries of the tropical and sub-tropical areas of the world, Plant et al. (1996) emphasized the need for high resolution geochemical data in such countries. Dissanayake and Weerasooriya (1986) produced one of the first hydrogeochemical atlases of Asia and they demarcated areas where the groundwater fluoride concentrations are exceedingly high. This map proved to be particularly useful in site investigations for the deep well programme in the dry zone of Sri Lanka where endemic fluorosis incidences are alarmingly high. Similar maps in other countries, most notably in China, have also had a variety of uses particularly when geochemical diseases are prevalent (e.g. maps of Se, F in China, Figures 12.1 and 12.2, respectively). Some developing countries have recognized the great value of geochemical maps and as stated by Reedman et al. (1996), there should be greater priority given to such projects within their national programmes.

GEOCHEMICAL MAPPING - CHINA'S EXAMPLE

China has over the last 25 years, made truly remarkable progress and achievements in the field of geochemical mapping and mineral exploration. It has obtained regional geochemical data for 39 elements from several millions of stream sediment samples covering an area of more than 6 million km^2 (Xie et al., 1997, 2004). Such a vast programme has obviously generated massive quantities of useful geochemical data which could be used not only for mineral exploration, but for medical geology as well. China is therefore one of the countries presently leading in research on medical geology. Production of maps in China showing endemic fluorosis, endemic goitre, selenium responsive Keshan and Kaschin-Beck diseases etc have been greatly benefited by geochemical mapping programs. Figures 12.1 and 12.2 show the geochemical maps of fluoride and selenium distribution, respectively in flood plain sediments of Chain.

Recently Xie et al. (2004) developed the concept of *"geochemical blocks"* which has revolutionized geochemical mapping and mineral exploration in China. More than 80% of new mineral discoveries in China during the last 2 decades were due to information provided by China's National Geochemical Mapping (Regional Geochemistry-National Reconnaissance, RGNR) project which was initiated in 1979 (Xie et al., 1981; Xie and Zheng, 1983). The study of *"hierarchy of nested geochemical patterns"* for tungsten across the whole of China (Xie and Yin, 1993) demonstrated that there are very broad geochemical patterns above local geochemical dispersion halos, trains, and fans surrounding ore deposits, which had been the main topic of exploration geochemistry (Hawkes and Webb, 1962).

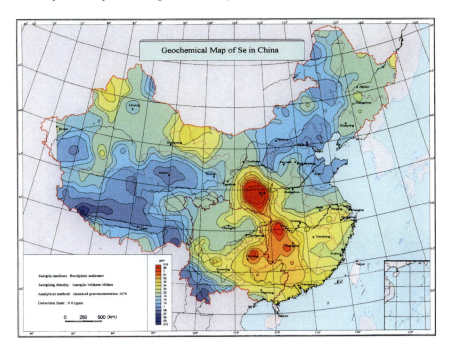

Fig. 12.1. Geochemical map of Se in China constructed using flood plain sediments (Courtesy of Prof. Xie Xuejing)

Table 12.1 shows the hierarchy of nested geochemical patterns and geochemical blocks. The concept of "geochemical block" with further modification has the potential of major application in delineating global "risk-areas" as affected by deficiency or toxicity of certain elements. Vast global areas such as those covered by the "goitre belt" spanning across Pakistan, Himalaya region, Bengal and Bangladesh, Vietnam and Indonesia could for example, be scientifically studied for their trace element behaviour using such geochemical concepts.

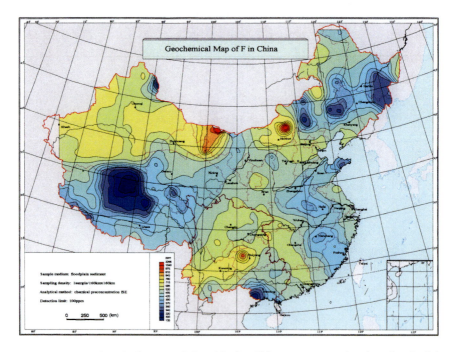

Fig. 12.2. Geochemical map of fluoride in China constructed using flood plain sediments (Courtesy of Prof. Xie Xuejing)

Table 12.1. Hierarchy of nested geochemical patterns and geochemical block ([a] The existence of geochemical continents can only be verified after wide-spaced geochemical mapping covered most part of the earth's land surface) (Xie et al., 2004)

Area (km^2)	Geochemical Patterns	
<100	Local anomaly	
100-1000	Regional anomaly (Geochemical region)	
1000-10,000	Geochemical province	
10,000-100,000	Geochemical mega-province	Surface expression of Geochemical block (given a 1000 m thickness)
100,000-1,000,000	Geochemical domain	
>1,000,000	Geochemical continent [a]	

SOIL MICRONUTRIENT MAPS IN TROPICAL COUNTRIES AND MEDICAL GEOLOGY

The importance of rock-soil-water-plant-human health relationships forms one of the basic factors influencing medical geology. As discussed in the earlier chapters, the chemical speciation of micronutrients and their bioavailability play a very significant role in the aetiology of geographically distributed diseases. A thorough knowledge of soil micronutrients of disease prone areas is therefore a necessary prerequisite for good epidemiological studies of regional medical studies. Micronutrient-poor soils are widespread throughout the world, most notably in the tropics and most food crops which form the staple diet of millions of people in the tropics are highly sensitive to the bioavailability of the micronutrients. Regional mapping of soil micronutrients, particularly in epidemiologically relevant disease-prone areas, is proving to be a most valuable exercise (White and Zasoski, 1999). Even though direct correlations between soil micronutrients and human health are rare, in tropical countries where most people live on home-grown food, such maps will have a variety of uses in agriculture, livestock, and human health.

Zinc is very well known as an important micronutrients and published maps of micronutrient-deficient areas in China indicated that about one-third of China's vast area was zinc-deficient (Zheng, 1991). These maps also showed areas of similar extent but different distributions that were low in B, Mo, and Mg while low copper areas represented only about 10% of the country (White and Zasoski, 1999).

The parent geological materials, paedogenic factors, geochemical environment, morphology and climate play different roles in the distribution of the micronutrients and there is, therefore a noticeable variation within and between regions.

Several methods have been used to produce maps depicting soil micronutrient content, plant-available micronutrients and areas of deficiency or toxicity at various cartographic scales. As reviewed by White and Zasoski (1999), small scale maps of global, continental or national extent are beneficial for a generalized overview of micronutrient deficiencies affecting livestock and human health. The micronutrient map of tropical Latin America was produced by Leon et al. (1985), and Kang and Osiname (1985) produced a similar map for tropical Africa. Among the problems observed in tropical Africa were (a) low total and extractable B, Cu, Mo and Zn in soils (b) local Mg deficiencies (c) iron toxicity in flooded rice soils. The parent geological materials were of more significance in the deficiency of the above elements in the soils.

Geochemical surveys of rocks, soils, vegetation, stream and lake sediments, surface and groundwaters, though used for purposes of mineral exploration have, however, yielded valuable information on micronutrients as well (Shacklette and Boerngen, 1984). Geochemical patterns of micronutrients in soils have been the subject of general studies (Thornton, 1987; 1993; Licht and Tarvainen, 1996; Plant et al., 2000). Recent advances in geographic information systems (GIS) and computer technology have facilitated the production of very informative geochemical maps depicting the status of micronutrients in the physical environment. Even though it is difficult to prove causality as against correlation, many attempts have been made to identify relationships between the geochemical environment and the incidence of diseases such as cancer, cardiovascular diseases, urinary stone formation. In certain lands such as Maputaland, South Africa, a tropical landmass covering an area of 50 x 100 km, the deficiency of micronutrients is extreme (Ceruti et al., 2003), and many diseases such as endemic osteoarthritis and dwarfism are prevalent. Geochemical maps of soil micronutrients will undoubtedly be of great value in such nutrient- impoverished lands.

FUTURE PROSPECTS FOR MEDICAL GEOLOGY

Medical geology is an emerging science and it is almost certain to achieve the status of an established field of science within a decade. Modern epidemiology requires a thorough understanding of the geochemical processes and pathways of essential and toxic elements and a clear integration of these different disciplines is being established.

Geochemical data bases such as those established, e.g., by the Global Geochemical Baseline Program of the IUGS/IAGC, British Geological Survey, Global Reference Network (GRN), Chinas' RGNR program. all provide valuable data useful in medical geochemistry. With recent advances in mapping, statistical techniques, extremely fast and accurate analytical techniques, the analysis of hundreds of samples rapidly, GIS based map making the possibilities for a better understanding of the aetiology of geochemically linked diseases, notably in the tropics is showing great promise.

Xie and Cheng (2001) have, for example, suggested that strategic deep penetrating NAMEG (NAnoscale Metals in Earth Gas) and MOMEO (MObile forms of MEtals in Overburden) geochemical methods developed in China are the optimum methods in geochemical mapping of deeply weathered lateritic terrains, so typical of tropical lands (Butt et al., 2000). The NAMEG method measures the nanoscale metal content carried out by micro-bubbles of "earth gas" generated form the interior of the earth. The MOMEO method measures the nanoscale content of mobile metals retained in soil when the carrier "earth gas" escapes into the atmosphere. Hidden ore deposits have been located using these techniques and medical geochemistry can be effectively integrated in such geochemical exploration programs.

Unlike in the past where only geological materials were analyzed and studied, it is now well established that for "terrain medical geochemistry" to be of use to the public and the health authorities, all samples from the area, i.e. rocks, minerals, soils, water, plants and food need to be studied together and correlated with in vivo studies. Such an approach has been taken by the British Geological Survey in studies on medical geochemistry.

In the medical field too, advances being made in the physiology of the human body, biochemical mechanisms of trace element ingestion, absorption and rejection, cell biology and inter-cellular chemical mechanisms may point to a greater need for the understanding of trace element geochemistry. The role of inorganic minerals such as apatite in the human bones and teeth is, for example, an area of research of extreme interest and which calls for a co-ordinated effort by geochemists, mineralogists and experts in human physiology. Among the other areas of interest are the iodine deficiency disorders (IDD), trace element-enzyme related diseases, natural radiation and mineral particles in the air.

"The human body is only a small part of a larger geochemical cycle. Medicine stands to gain by the proper understanding and application of Geology"

References

Abedin MJ, Feldman J, Meharg AA (2002) Uptake kinetics of arsenic species in rice plants. Plant Physiology 128 (3): 1-9

Aber JD, Nadelhoffer KJ, Steudler P, Melillo JM (1989) Nitrogen saturation in northern forest ecosystems. Bioscience 39: 378-386

Abrahams PW (1997) Geophagy (soil consumption) and iron supplementation in Uganda. Trop Med Int Health 2: 617-623

Abrahams PW (1999) The chemistry and mineralogy of three savannah lick soils. J Chem Ecol 25: 2215-2228

Abrahams PW, Parsons JA (1996) Geophagy in the tropics: A literature review. Geog. J 162: 63-72

Abrahams PW, Parsons JA (1997) Geophagy in the tropics: An appraisal of three geophagical materials. Environmental Geochemistry and Health 19: 19-22

Achryya SK, Lahiri S, Raymahashmy BC, Bhowink A (2000) Arsenic toxicity of groundwater in parts of Bengal basin in India and Bangladesh the role of Quaternary stratigraphy and Holocene sea-level fluctuation. Environmental Geology 39: 1127-1137

Adetunji MT (1993) Nitrogen application and under groundwater contamination in some agricultural soils of south western Nigeria. Fertilizer Research 37: 159-163

Adriano D (2001) Trace elements in terrestrial environments. 2nd edition Springer, New York, Berlin, Heidelberg. 867 pp

Agget PJ, Mills CF, Morrison A, Callan M, Plant J, Simpson PR, Stevenson A, Dingwall-Fordyce I, Halliday CF (1988) A study of environmental geochemistry and health in northeast Scotland. In: Thornton I (ed) Proc 2nd Int Symp on Geochemistry and health, Northwood, p 81-91

Ahman D, Krumholz LR, Hemond H, Lovely DR, Morel FMM (1994) Microbe grows by reducing arsenic. Nature 371: 750

Ahman D, Krumbolz LR, Hemond HF, Lovely DR, Morel FMM (1997) Microbial mobilization of arsenic from sediments of the Aberjona Watershed. Environ Sci Tech 31: 2923-2930

Ahmed KM, Hoque M, Hasan MK, Ravenscroft P, Chowdhury LR (1998) Occurrence and origin of water well methane gas in Bangladesh. J Geol Soc India 51: 697-708

Ahn HW, Fulton B, Moxon D, Jeffery EH (1995) Interactive effects of fluoride and aluminium uptake and accumulation in bones of rabbits administered both agents in their drinking water. J Toxicol Environ Health 44: 337-350

Alam MK, Hasan AKMS, Khan MR, Whitney JW (1990) Geological Map of Bangladesh. Scale 1:1000000, Geol Survey Bangladesh

Allaway WH (1968) Controls on the environmental levels of selenium. In: Hemphill DD (ed) Trace substances in Environmental Health, p 181-206

Allison FE (1966) The fate of nitrogen applied to soils. Adv Agron 18: 219-258

Anonymous (2002) Fluorides. Environ Health Criteria 227: 231-23

Apambire WB, Boyle DR, Michel FA (1997) Geochemistry, genesis and health implications of fluoriferous groundwaters in the upper regions of Ghana. Environmental Geology 33: 13-24

Aposhian HV, Gurzau ES, Le XC, Gurzau A (2000) Occurrence of monomethylarsonous acid in urine of humans exposed to inorganic arsenic. Chem Res Toxicol 13: 693-697

Appelo CAJ, Postma D (1994) Geochemistry groundwater and pollution. AA Balkema-Rotterdam-Brookfield, 536 pp

Arthur JR, Beckett CT (1994) New metabolic roles for selenium. Proc Nutr Soc 53: 615-624

Arthur JR, Beckett CT, Mitchell JH (1999) The interaction between selenium and iodine deficiencies in man and animals. Nutr Res Rev 12(1): 55-73

Aswathanarayana U, Lahermo P, Malisa E, Nanyaro JT (1985) High fluoride waters in an endemic fluorosis area in northern Tanzania. In: Thornton I (ed) Proceedings of the 1st International Symposium on Geochemistry and Health Monograph Series: Environmental Geochemistry and Health, p 243 -249

Aufreiter S, Hancock RGV, Mahaney WC, Stambolic-Robb A, Sanmugadas K (1997) Geochemistry and mineralogy of soils eaten by humans. Int J Food Sci Nutrit 48: 293–305

Australian Institute of Geoscientists (2003) Angela Giblin Groundwater-Pathfinders to concealed are deposits. AIG Newsletter No 71

Australian Radiation Protection and Nuclear Safety Agency (ARPANSA) (2004) Radiation basics - Health effects of ionizing radiation, Fact Sheet 17, 3pp.

Azzam EI, Raaphorst GP, Mitchel RE (1994) Radiation-induced adaptive response for protection against micronucleus formation and neoplastic transformation in C3H 10T1/2 mouse embryo cells. Radiation Res 138: 528-531

Bailey JC (1977) Fluorine in granitic rocks and melts- a review. Chem Geol 19:1-42

Balasuriya S, Perera PAJ, Herath KB, Katugampola SL, Fernando MA (1992) Role of iodine content of drinking water in the aetiology of goitre in Sri Lanka. Ceylon J Medical Science 35: 45-51

Banzi F, Kifanga LD, Bundala FM (2000) Natural radioactivity and radiation exposure at the Minjingu phosphate mine in Tanzania. J Radiol Prot 20: 41-51

Barrow NJ, Ellis AS (1986) Testing a mechanistic model III -The effects of pH on fluoride retention by a soil. J Soil Sci 37:287–293

Bachou H (2002) The nutrition situation in Uganda. Nutrition 18 (4): 356-358

Baum MK, Shor-Posner G, Lai S, Zhang G, Lai H, Fletcher MA, Sauberlich H, Page JB (1997) High risk of HIV-related mortality is associated with selenium deficiency. J Acquir Immune Defic Syndr Hum Retro 15: 374

Bautista EM, Alexander M (1972) Reduction of inorganic compounds by soil microorganisms. Soil Sci Soc Am Proc 36: 918-920

Behrens H (1982) New insights into the chemical behaviour of radioiodine in aquatic environments. In: Environmental migration of long-lived radionuclides International Atomic Energy Agency, Vienna, p 27–40

Bellizia V, De Nickola L, Minutolo R, Russo D, Cianciarieso B, Andrencci M, Conte G, Andreucci VE (1999) Effects of water hardness on urinary risk factors for kidney stones in patients with idiopathic nephrolithiasis. Nephron 81 Suppl 1: 66-70

Berger AR (1999) Natural chemical hazards to human health - towards an interdisciplinary research agenda. Paper read at IGU Workshop on Setting an Agenda for Research on Health and Environment: Health and Environmental Resources Mona Jamaica Nov 12-14 1999

Bernadi D, Dini FL, Azzrelli A, Giaconi A, Volterrani C, Lunardi M (1995) Sudden cardiac death rate in an area characterized by high incidence of coronary artery disease and low hardness of drinking water. Angiology 46: 145-149

Bettencourt AO, Teixeria MM, Elias MD, Madruga MJ (1990) Environmental monitoring in uranium mining areas. The Environmental Behaviour of Radium (IAEA Technical Report Series 310), Vienna, p 281–294

Bhattacharya P, Chatterjee D, Jacks G (1997) Occurrence of arsenic-contaminated groundwater in alluvial aquifers from the Delta Plain Eastern India: options for a safe drinking water supply. Water Res Dev 13: 79-92

Bianconi F (1987) Uranium geology of Tanzania. Monograph series on mineral deposits 27: 11-25

Bichler KH, Eipper E, Naber K, Braun V, Zimmerman R, Lahme S (2002) Urinary infection stones. Int J Antimicrobial Agents 19: 488-498

Binh HV, Lap NB, Thang TT (1992) Iodine geochemistry and goitre-cretinism in Southeast Asia. In Regional Seminar on Environmental Geology, 11-13 Nov 1992, Hanoi

Bleichrodt N, Born M (1994) A meta-analysis of research on iodine and its relationship to cognitive development. In: Stanbury JB (ed) The damaged brain of iodine deficiency New York Cognisant Communication Corporation, p 195-210

Blundell G, Henderson WJ, Price EW (1989) Soil particles in the tissues of the foot in endemic elephantiasis of the lower legs. Ann Trop Med Parasit 83: 381-385

Bolton KA, Campbell VM, Burton FD (1998) Chemical analysis of soils of Kowloon (Hong Kong) eaten by hybrid Macques. J Chem Ecol 24: 195-205

Bourdoux P, Delange F, Gerard M, Mafuta M, Hanson A, Ermans AM (1978) Evidence that cassava ingestion increases thiocyanate formation: a possible etiologic factor in endemic goitre. J Clin Endocrinol Metab 46(4): 613-621

Bowie SHU, Thornton I (eds) (1985) Environmental Geochemistry and health. Reidel Publishing Co., Dordrecht, The Netherlands, 140 pp

Bowman WD (2000) Biotic controls over ecosystem response to environmental change in alpine tundra of the Rocky Mountains. Ambio 29: 396-400

Boyle RW (1974) Elemental associations in mineral deposits and indicator elements of interest in geochemical prospecting (revised edition). Geol Surv Canada Prof Paper, Ottawa, 40 pp

Braman RS (1975) Arsenic in the environment In: Arsenical Pesticides. In: EA Woolson (ed), ACS Symp Ser, Washington DC, American Chemical Society

Brammer A (1996) The Geography of Soils of Bangladesh. University Press Ltd Dhaka

Braun JJ, Dupre B, Viers J, Ngoupayou JRN, Bedimo JPB, Nkamdjou LS, Freydier R, Robain H, Nyeck B, Bodin J, Oliva P, Boeglin J-L, Stemmler S, Berthelin J (2002) Biogeohydrodynamic in the forested humid tropical environment: the case study of the Nsimi small experimental watershed (south Cameroon). Bulletin de la Société Géologique de France 173 (4): 347-358

Brewer RF (1966) Fluorine. In: Chapman HD (ed) Diagnostic criteria for plants and soils Riverdale California. University of California Division of Agricultural Science, p 180-195

Brindley GW (1978) The structure and chemistry of hydrous nickel-containing silicate and aluminium minerals. Bull BRGM II: 3 233-245

Brindley GW, Hang PT (1973) The nature and of garnierites-I Structures chemical compositions and colour characteristics. Clay and Clay minerals 21(1): 27-40

Brindley GW, Maksimovic Z (1974) The nature and nomenclature of hydrous nickel-containing silicates. Clay Mineral Bull 10: 271-277

Brindley GW, Wan HM (1975) Compositions structures and thermal behaviour of nickel-containing minerals in the lizardite-nepouiute series. Am Mineralogist 60: 863-871

British Geological Survey (2001) Arsenic Contamination of Groundwater in Bangladesh. BGS/DFID Technical Report WC/00/19, Keyworth UK

Broadbent FE, Rauschkolb RS (1977) Nitrogen fertilization and water pollution. Calif Agric 31: 24-25

Brouwer ID, Dirks OB, De Bruin A, Hautvast JG (1988) Unsuitability of world health organization guidelines for fluoride concentrations in drinking water in Senegal. Lancet 33 (1): 223-225

Brown B, Neff J (1993) Bioavailability of sediment-bound contaminants to marine organisms. Report PNL-8761 UC-0000 Prepared by Batelle Marine Sciences Laboratory for the National Ocean Pollution Programme Office NOAA

Bruning-Fann CS, Kaneene JB (1993a) The effects of nitrate, nitrite and N-nitroso compounds on human health; a review. Vet Hum Toxicol 35(5): 521-538

Bruning-Fann CS, Kareene JB (1993b) The effects of nitrate, nitrite and N-nitroso compounds on animal health. Vet Hum Toxicol 35(3): 237-253

Bureau of Indian Standards (1991) Specification for Drinking Water (BIS: 10500) New Delhi, India

Butt M, Lintern MJ, Anand RR (2000) Evolution of regoliths and landscapes in deep weathered terrain-implications for geochemical exploration. Ore Geology Reviews 16: 167-183

Campa A, Shor-Posner G, Indacochea F, Zhang G, Lai H, Asthana D, Scott GB, Baum MK (1999) Mortality risk in selenium-deficient HIV-positive children.

J of Acquired Immune Deficiency Syndromes and Human Retrovirology 20 (5): 508-513

Cannon WB (1932) The wisdom of the body. WW Norton & Co, New York

Cao J, Zhao Y, Liu J, Xirao R, Danzeng S, Daji D, Yan Y (2003) Brick tea fluoride as a main source of adult fluorosis. Food and Chemical Toxicology 41(4): 535-542

Carrol MF, Schode DS (2003) A practical application to hypercaleemia. Am Family Physician 67: 1959-1966

Cave B, Kolsky P (1999) Groundwater latrines and health. London School of Hygiene and Tropical Medicine UK and WEDC Loughborough Univ, 30 pp

Cerklewski FL (1997a) Fluoride bioavailability-nutritional and clinical aspects. Nutrition Research 17(5): 907-929

Cerklewski FL (1997b) Fluorine. In: O'Dell BL Sunda RA (eds) Handbook of nutritionally essential mineral elements. Marcel Dekker New York. p 583-602

Ceruti P, Fey M, Pooley J (2003) Soil nutrient deficiencies in an area of endemic osteoarthritis (Mseleni joint disease) and dwarfism in Maputaland South Africa. In: Skinner HC, Berger AR (eds) Geology and Health Closing the Gap Oxford University Press, p 151-154

Chapin SF, Matson PA, Mooney HA (2002) Principles of Terrestrial Ecosystem Ecology. Springer, Heidelberg, Berlin

Chatin A (1852) Recherche de l'iode dans l'air les eaux, le sol et la France et du Piémout Compl Rend Acad Sci (Parise) 34:51-54

Chebotarev II (1955) Metamorphism of natural waters in the crust of weathering. Geochim Cosmochim Acta 8 (3):137-170

Chernet T, Travi Y (1993) Preliminary observations concerning the genesis of high fluoride contents in the Ethiopian Rift. In: Thorweihe U, Schandelmeier H (eds) Geoscientific Research in the Northeast Africa, Balkema Rotterdam, p 651-655

Chernet T, Travi Y, Valles V (2001) The mechanism of degradation of the quality of natural water in the lakes region of the Ethiopian Rift Valley. Water Research 35(12): 2819-2832

Choubisa SL (2001) Endemic fluorosis in southern Rajasthan India. Fluoride 34(1): 61-70

Chukrov EV (1975) Hypergene iron oxides in geological processes (in Russian). Acad Sci USSR Publ House Nauka 170 pp

Chowdary VM, Rao NH, Sarma PBS (2004) A coupled soil water and nitrogen balance model for flooded rice fields in India. Agric Ecosyst Environ 103(3): 425–441

Christensen H, Dharmagunawardena HA (1986) Hydrogeological investigations in hard rock terrains of Sri Lanka with special emphasis on Matale and Polonnaruwa Districts. In: Proc on groundwater and water quality in Sri Lanka Institute of Fundamental studies Sri Lanka, 25th Oct

Christensen H, Dharmagunawardhena HA (1987) Behaviour of some chemical parameters in tube well water in Matale and Polonnaruwa districts. In: Dissanayake CB, Gunatilake AAL (eds) Some aspects of the chemistry of the en-

vironment of Sri Lanka Sri Lanka, Association for Advancement of Science Colombo, p 45-75
Churchill DN, Bryant D, Fodor G, Gault MH (1978) Drinking water hardness and urolithiasis. Ann Intern Med 88: 513-514
Churchill DN, Maloney CM, Bear J, Bryant DG, Fodor G, Gault MH (1980) Urolithiasis-a study of drinking water hardness and genetic factors. J Chronic Dis 33: 727-731
Cisar JO, Xu DQ, Thompson J, Swaim W, Hu L, Kopecko DJ (2000) An alternative interpretation of nanobacteria-induced biomineralization. Proc Nat Acad Sci USA 97(21): 11511-11515
Clancy J, Mc Vicar A (1995) Physiology and anatomy: a homeostatic approach. Edward Arnold London, 734 pp
Clarke MCG, Woodhall DG, Allen D, Darling G (1990) Geological Volcanological and hydrological controls on the occurrence of geothermal activity in the area surrounding Lake Naivasha Kenya. Ministry of Energy Report Nairobi Kenya, 138 pp
Colquhon J (1990) Flawed foundation: a re-examination of the scientific basis for a dental benefit from fluoridation. Community Health Studies 14(3): 288-296
Comly HH (1945) Cyanosis in infants caused by nitrates in well water. J Am Med Assoc 129: 112-116
Comstock GW (1979) Water hardness and cardiovascular diseases. American Journal of Epidemiology 110: 375-400
Cook TD, Bruland KW (1987) Aquatic Chemistry of Selenium Evidence of biomethylation. Environ Sci Technol 21: 1214-1219
Cooksey RC, Gaitan E, Lindsay RH, Hill JB, Kelley K (1985) Humic substances a possible source of environmental goitrogens. Organic Geochemistry 8(1): 77-80
Cooray PG (1978) Geology of Sri Lanka. 3rd Regional Conf of Geological Mineral Resources of SE Asia, Bangkok, p 14-78
Corliss W (1999) Anomalies in Geology: Physical, Chemical, Biological Science Frontiers, MD, USA, 335pp
Correns CW (1956) Geochemistry of halogens. Phys Chem Earth 1: 161-182
Cowan A (1976) A natural fission reactor. Scientific American 235: 36-47
Crawford MD, Clayton DG, Stanley F, Shaper AG (1977) An epidemiological study of sudden death in hard and soft water areas. J Chron Diseases 30: 69-80
Cronin SJ, Neall JE, Lecointre MA, Hedley P, Loganathan P (2003) Environmental hazards of fluoride in volcanic ash - a case study from Ruapehu Volcano New Zealand. J Volcano & Geotherm Res 121: 271-291
Cullen WR, Reimer KJ (1989) Arsenic speciation in the environment. Chem Rev 89: 713-764
Cummings DE, Caccavo F, Fendorf S, Rosenzweig RF (1999) Arsenic mobilization by the dissimilatory Fe (III) reducing bacterium *Shewanella alga* BrY. Environ Sci Technol 33: 723 -729
Dai S, Ren D, Ma S (2004) The cause of endemic fluorosis in western Guizhon Province Southwest China. Fuel 83: 2095-2098

Darnley AG (1995) International geochemistry mapping- a review. J Geochem Explor 55: 5-10

Darrouzet-Nardi A (2005) Remote sensing of earth's nitrogen cycle. Class paper University of Colorado at Boulder, Department of Ecology and Evolutionary Biology and the Institute of Arctic and Alpine Research

Das D, Samanta G, Mandal BK, Chowdhury TR, Chanda CR, Chowdhury PP, Basu GK, Chakraborti D (1996) Arsenic in groundwater in six districts of West Bengal India. Environ Geochem Health 18: 5-15

Davies TC (1994) Water quality characteristics associated with fluorite mining in the Keno Valley area of western Kenya. Int J Health Research 4: 165-175

Davies TC (1996) Geomedicine in Kenya. J African Earth Sciences 23(4): 577-591

Davis SN, RJM DeWiest (1966) Hydrogeology. John Wiley and Sons. New York, NY, 463 pp

De SK, Rao SS, Tripathi CM, Rai C (1971) Retention of iodine by soils clays. Indian J Agri Chem 40: 43-49

Dean H, McKay F (1939) Production of mottled enamel halt by a change in common water supply. Am J Public Health 29: 590

Dean HT (1942) The investigation of physiological effects by the epidemiological method. In: Moulton FR (ed) Fluorine and dental health Washington DC, American Association for the Advancement of Science 19: 23-31

Delange F, Bourdoux P, Camus M, Gerard M, Mafuta M, Hanson A, Ermans AM (1976) The toxic effect of cassava on human thyroid. In: Proceedings of the 4th Symposium of the International Society for Tropical Food Crops (IDRC Ottawa Canada), p 237–242

Derbyshire E (2003) Natural dust and pneumoconiosis in high Asia. In: Skinner HCW, Berger AR (eds) Geology and Health: closing the gaps, Oxford University Press New York, p 15-18

Deshmukh AN, Wadaskar PM, Malpe DB (1995) Fluorine in environment: A review. In: Deshmukh AN, Yedekar DB, Nair KKK (eds) Fluorine in Environment Gondwana, Geological Magazine Special Issue 9: 1–20

Deverel SJ, Gillion RJ, Fujii R, Izbicki JA, Fields JC (1984) Areal distribution of selenium and other inorganic constituents in shallow ground water of the San Luis Drain Service Area San Joaquin Valley California, A preliminary study. US Geol Surv Water-Resour Invest Rep No 84-4319 Sacramento CA, 67 pp

De Vletter DR (1978) Criteria and problems in estimating global lateritic nickel resources. Mathematical Geology 10:533-542

Diamond J, Bishop KD, Gilardi JD (1999) Geophagy New Guinea birds. Ibis 141: 181-193

Dias da Cunha, Leite CVB, Zays Z (2004) Exposure to mineral sands dust particles-Nuclear Instruments and Methods in Physics Research. Section B: Beam Interactions with Materials and Atoms 217 (4): 649-656

Diplock AT (1994) Antioxidant and disease prevention. Mod Aspects Medicine 15: 293-376

Dissanayake CB (1984a) Weathering of nickeliferous serpentinites in humid tropical terrains. In: Roy S. Ghosh SK (eds) Products and Processes of Rock

Weathering: Recent Researches in Geology, Hindustan Publishing Corporation India, p 1-15

Dissanayake CB (1984b) Environmental geochemistry and its impact on humans. In: Fernando CH (ed) Ecology and Biogeography of Sri Lanka: Monographiae Biologicae Junk Publishers, The Netherlands, p 65-97

Dissanayake CB (1991a) The fluoride problem in the groundwater of Sri Lanka-environmental management and health. Intl J Environ Studies 38: 195-203

Dissanayake CB (1991b) Humic substances and chemical speciation: implications on environmental geochemistry and health. Int J Environ Studies 37: 247-258

Dissanayake CB (1996) Water quality and dental health in the Dry Zone of Sri Lanka. In: Appleton JD, Fuge R, McCall GJH (eds) Environmental geochemistry and health. Geological Society UK Special Publication 113, p 131–141

Dissanayake CB, Chandrajith RLR (1993) Geochemistry of endemic goitre Sri Lanka. Applied Geochemistry Suppl Issue No 2: 211-213

Dissanayake CB, Chandrajith R, Tobschall HJ (1998) The iodine cycle in the tropical environment-implications on iodine deficiency disorders. Int J Environ Studies 56: 357-372

Dissanayake CB, Chandrajith R (1999) Medical geochemistry of tropical environments. Earth Science Reviews 47: 219-258

Dissanayake CB, Chandrajith R (2006) Inorganic aspects of medical geology. Z dt Ges Geowiss 157 (3): 9-18

Dissanayake CB, Weerasooriya SVR (1986) The Hydrogeochemical Atlas of Sri Lanka Natural Resources Energy and Science Authority of Sri Lanka Colombo, 103 pp

Dissanayake CB, Weerasooriya SVR (1987) Medical geochemistry of nitrates and human cancer in Sri Lanka. Intl J Environ Studies 30: 145-156

Dissanyake CB, Van Riel (1976) A recently discovered nickeliferous serpentinite from Uda Walawe Sri Lanka. Geol Mijnbouw 57: 91-92

Dissanayake JK, Abeygunasekara A, Jayasekara R, Ratnatunga C, Ratnatunga NVI (1994) Skeletal fluorosis with neurological complications (case report). Ceylon Med J 39: 48-50

Djazuli M, Bradbury JH (1999) Cyanogen content of cassava roots and flour in Indonesia. Food Chemistry 65: 523-525

Doran JW (1982) Microorganisms and the biological cycling of selenium. In: Marshall KC (ed) Advances in Microbial Ecology. Plenum Press NY, p 1-32

DPHE (1999) Groundwater studies for arsenic contamination in Bangladesh. Final Report-Rapid Investigation Phase. Department of Public Health Engineering Government of Bangladesh Mott MacDonald and Brit Geol Survey

Driscoll KE, Maurer JK (1991) Cytokine and growth factor release by alveolar macrophages potential biomarkers of pulmonary toxicity. Toxicol Pathol 19: 398-405

Dungan RS, Frankenberger Jr WT (1999) Microbial transformations of selenium and the bioremediation of seleniferous environments. Bioremediation Journal 3(3):171-188

References

Dunn JT, van der Haar F (1990) A practical guide to the correction of iodine deficiency. International Council for Control of Iodine Deficiency Disorders. Technical Manual 3, UNICEF/WHO/ICCIDD, The Netherlands

Durlach J, Bara M, Guiet-Bara A (1985) Magnesium level in drinking water and cardiovascular risk factor: a hypothesis. Magnesium 4: 5-15.

Durlach J, Bara M, Guiet-Bara A (1989) Magnesium level in drinking water: its importance in cardiovascular risk. In: Itokawa Y, Durlach J (eds) Magnesium in health and disease, John Libbey London, p 173-182

Dzombak DA, Morel FMM (1990) Surface Complexation Modelling- Hydrous Ferric Oxide. John Wiley New York

Eapen J (1998) Elevated levels of cerium in tubers from regions endemic for endomyocardial fibrosis (EMF). Bull Environ Contaminant Toxicol 60: 168-170

Edmunds WM (1996) Indicators in the groundwater environment of rapid environmental change In: Berger AR, Iams WJ (eds) Geoindicators: assessing rapid environmental changes in earth systems, Balkema Rotterdam, p 135–150

Edmunds WM, Smedley PL (1996) Groundwater geochemistry and health: an overview. In: Appleton JD, Fuge R, McCall GJH (eds) Environmental Geochemistry and Health, Geological Society Special Publication UK 113: p 91-95

Edmunds WM, Smedley PL (2004) Fluoride in natural waters. In: O Selinus (ed) Essentials of Medical Geology. Elsevier, p 301-329

Eisenberg MJ (1992) Magnesium deficiency and sudden death. Am Heart J 124: 544-549

Eisenbud M (1973) Environmental radioactivity. Academic Press NY, 542 pp

Ekpechi OL (1967) Pathogenesis of endemic goitre in eastern Nigeria. Br J Nutr 21: 537-545

Ekstrom TK (1972) The distribution of fluorine among co-existing minerals. Contr Min and Pet 34: 192-200

Ekstand J (1978) Relationship between fluoride in the drinking water and the plasma fluoride concentration in man. Caries Research 12: 123-127

Esson J, Carlos L (1978) The occurrence mineralogy and chemistry of some garnierites from Brazil. Bull BRGM II: 263-274

FAI (Fertilizer Association on India) (1997) Fertilizer statistics- IV. New Delhi 75

FAO (Food and Agriculture Organization) (1978) Food and Agriculture production year book. United Nations Rome

FAO (1991) Fertilizer yearbook-1990. United Nations, Rome, Stat Ser 99, vol 40

FAO (1993) FAO production year book. United Nations Rome, Stat Ser 100, vol 44

Faust GT (1966) The hydrous nickel-magnesium silicates- the garnierite group. Am Mineralogist 51: 279-298

Federman DG, Kirsner RS, Federman GSL (1997) Pica. Are you hungry for facts? Conn Med 61: 207-209

Fergeson JL, (2002) Endomyocardial fibrosis. eMedicine http://www.emedicinecom/med/topic677.htm (assed on 10 November 2006)

Fergusson JE (1990) The Heavy elements Chemistry- Environmental Impact and Health, Pergamon Oxford, 614 pp

Fernando MA, Balasuriya S, Herath KB, Katugampola S (1987) Endemic goitre in Sri Lanka. In: Dissanayake CB, Gunatilaka AAL (eds) Some aspects of the Chemistry of the environment of Sri Lanka. Association for the Advancement of Science Colombo, p 45-65

Fernando MA, Balasuriya S, Herath KB, Katugampola S (1989) Endemic goitre in Sri Lanka. Asia-Pacific Journal of Public Health 3(1): 11-18

Fishbein L (1979) Overview of some aspects of occurrence formation and analysis of nitrosamines. Sci Total Environ 13 (2): 157-188

Fisher RB, Dressel WM (1959) The Nicaro (Cuba) nickel ores-basic studies including differential thermal analysis in controlled atmospheres. U.S. Bur Mines Report Inv 244: 650-657

Fleischer M, Robinson WO (1963) Some problems of geochemistry of fluoride. In: Shaw DM (ed) Studies in Analytical Geochemistry. Royal Society of Canada Special Publications No 6, University of Toronto Press Toronto, p 58-75

Flühler H, Polomski J, Blaser P (1982) Retention and movement of fluoride in soils. J Environ Qual 11: 461-468

Food and Nutrition Board Indian institute of Medicine (1997) Dietary Reference Intake (1999) Dietary Reference Intakes for Calcium Phosphorus Magnesium Vitamin D and Fluoride. Institute of Medicine, National Academy Press

Forbes GB (1990) Body composition. In: Brown ML (ed) Present knowledge in nutrition. 6th ed Washington DC ILSI Press, p 7-12

Fordyce FM, Masara D, Appleton JD (1996) Stream sediment soil and forage chemistry as indicators of cattle mineral status in northeast Zimbabwe In: Appleton JD, Fuge R, McCall GJH (eds) Environmental Geochemistry and health. Geological Society Special Publication UK 113: p 23-37

Fordyce F, Johnson CC, Navaratne URB, Appleton JD, Dissanayake CB (1998) Studies of selenium geochemistry and distribution in relation to iodine deficiency disorders in Sri Lanka. Tech Report WC/98/28, Overseas Geology Series BGS-UK

Fordyce F, Johnson CC, Navaratne URB, Appleton JD, Dissanayake CB (2000a) Selenium and iodine in soil rice and drinking water in relation to endemic goiter in Sri Lanka. Sci of Total Environ 263: 127-141

Fordyce FM, Guangdi Z, Green K, Xinping L (2000b) Soil grain and water chemistry in relation to human selenium-responsive diseases in Enshi District China. Applied Geochemistry 15 (1): 117-132

Fordyce FM, Stewart AG, Ge X, Jiang JY, Cave M (2002) Environmental controls in IDD A case study in the Xinjiang Province of China. BGS Technical Report CR/01/045N 130 pp

Förstner U, Whittman GTW (1981) Metal pollution in the aquatic environment. Springer, Berlin, 486 pp

Foster HD, Zhang L (1995) Longevity and selenium deficiency: evidence from the Peoples Republic of China. Sci Total Environ 170: 133-139

Fourie AB, van Ryneveld MB (1995) The fate in the subsurface of contaminants associated with on-site sanitation: a review. Water SA 21(2): 101-111

Freeze RA, JA Cherry (1979) Groundwater. Prentice Hall Inc, Englewood Cliffs, NJ, 604 pp

Fuge R (1996) Geochemistry of iodine in relation to iodine deficiency diseases. In: Appleton JD, Fuge R, Mc Call GJH (eds) Environmental Geochemistry and Health. Geol Soc Sp Publ UK, p 201-213

Fuge R, Johnson CC (1986) The geochemistry of iodine-a review. Env Geochem Health 8: 31-54

Fuhrmann M, Bajt S, Schoonen MAA (1998) Sorption of iodine on minerals investigated by X-ray absorption near edge structure (XANES) and ^{125}I tracer sorption experiments. Applied Geochemistry 13: 127-141

Fujinami N, Koga T, Morishima H (1999) Preliminary survey of absorbed dose rates in air at Guarapari and Meape in Brazil. Hoken Batsuri 34: 253-267

Fukagawa M, Kurokawa K (2002) Calcium homeostasis and imbalance. Nephron 92: 41-45

Furguson JF, Gavis J (1972) A review of the As cycle in natural waters. Water Research 6: 1259

Fyfe WS, Kronberg BI, Leonardos OH, Olorunfemi N (1983) Global tectonics and agriculture a geochemical perspective. Agric Ecosys Environ 9: 383-399

Gabor S, Anca Z, Zugravu E (1975) In vitro action of quartz on alveolar macrophage lipid peroxides. Archives of Environmental Health 30: 499-501

Gaciri SJ, Davies TC (1993) The Occurrence and geochemistry of fluoride in some natural waters of Kenya. J Hydrology 143(3/4): 395-412

Gaitan E (1990) Goitrogens in food and water, Annu Rev Nutr 10: 21-39

Galloway JN, Aber JD, Erisman JW, Seitzinger SP, Howarth RW, Cowling EB, Cosby BJ (2003) The nitrogen cascade. Bioscience 53: 341-356

Galloway JN, Dentener FJ, Capone DG, Boyer EW, Howarth RW, Seitzinger SP, Asner GP, Cleveland CC, Green PA, Holland EA, Karl DM, Michaels AF, Porter JH, Townsend AR, Vorosmarty CJ (2004) Nitrogen cycles: past, present, and future. Biogeochemistry 70:153-226

Gardner M (1973) Soft water and heart disease? In: Health and the Environment, Lenihan J and Fletcher WW (eds) Blackie, Glasgow & London 121

Garth D (2001) Hypokalemia (www.emedicine.com/EMERG/topic273.htm)

Geissler PW, Mwaniki DL, Thiong F, Friis H (1997) Geophagy among primary school children in Western Kenya. Tropical Medicine and International Health 2: 624-630

Geissler PW, Prince RJ, Levene M, Poda C, Beckerleg Mutemi W, Shulman CE (1999) Perceptions of soil-eating and anaemia among pregnant women on the Kenyan. Coast Social Science and Medicine 48: 1069-1079

Gelinas Y, Randall H, Robidoux L, Schmit JP (1996) Well water survey in the two districts of Conakry (Republic of Guinea) and comparison with the piped city water. Water Research 30(9): 2017-2026

Gembicki M, Herath KB, Piyasena RD, Wickremanayake TW (1973) Thyroid radioiodine uptake studies on some euthyroid Ceylonese. Ceylon Medical J 18: 134-137

Ghiassi-nejad M, Mortazavi SMJ, Cameron JR, Niroomand-rad A, Karam PA (2002) Very high background radiation areas of Ramsar Iran, Preliminary biological studies. Health Physics J 82: 87-93

Ghosh BC, Bhat R (1998) Environmental hazards of nitrogen loading in wetland rice fields. Environ Pollut 102: 123-126

Ghosh NC, Sharma CD, Sinha SN (1986) An appraisal of the quality of drinking water resource in Bihar-A case study form eastern India. Indian J Landscape System and Ecological Studies 9: 1

Gibson D (1974) Descriptive human pathological mineralogy. American Mineralogist 5a: 1177-1182

Gibson S (1992) Effects of fluoride on immune system function. Complimentary Med Res 63: 111-113

Gilardi JD, Duffey SS, Munn CA, Tell LA (1999) Biochemical functions of geophagy in parrots detoxification of dietary toxins and cytoprotective effects. J Chem Ecol 25: 897-922

Gillberg M (1964) Halogen and hydroxyl contents of micas and amphiboles in Swedish granite rocks. Geochim Cosmochim Acta 28: 495-516

Goldberg S, Johnston CT (2001) Mechanism of arsenic adsorption on amorphous oxides evaluated using macroscopic measurements vibrational spectroscopy and surface complexation modelling. J Colloid Interface Sci 234: 204-216

Goldberg S (2002) Competitive adsorption of arsenate and arsenite on oxides and clay minerals. Soil Sci Soc Am J 66: 413-421

Goldschmidt VM (1954) Geochemistry. In: A Muir (ed) Oxford University Press, London

Goraczko W (2000) Ionizing radiation and mitogenetic radiation: two links of the same energetic chain in a biological cell. Medical Hypotheses 54: 461-468

Gordon JJ, Quastel GH (1948) Effect of organic arsenicals on enzyme system. Biochem J 42: 337-350

Grant D (1986) Fluoride-The poison in our midst. Ecologist 16(6): 249-252

Greer MA, Langen P (1997) Antithyroid substances and naturally occurring goitrogens S Karger Basel 178 pp

Hamid A, Warkentin BP (1967) Retention of ^{131}I used as a tracer in water-movement studies. Soil Science 104: 279-282

Handa BK (1975) Geochemistry and genesis of fluoride containing groundwaters in India. Groundwater 13: 275-281

Halbert BE, Chambers DB, Cassaday VJ, Hoffman PO (1990) Environmental assessment modelling. The Environmental Behaviour of Radium. IAEA Technical Report Series 310, Vienna, pp 345-391

Harvey CF, Swartz CH, Badruzzaman ABM, Keon-Blute K, Yu W, Ashraf Ali M, Jay J, Beckie R, Niedan V, Brabender D, Oates PM, Ashfaque KN, Islam S, Hemond HF, Ahmed MF (2002) Arsenic mobility and groundwater extraction in Bangladesh. Science 298: 1602-1606

Harvey R, Powell JJ, Thompson RPH (1996) A review of the geochemical factors linked to podoconiosis. In: Appleton JD, Fuge R, Mc Call GJH (eds) Environmental Geochemistry and health. Geol Soc Sp Publ UK, 113, p 255-260

Hawkes HE, Webb JS (1962) Geochemistry in Mineral Exploration. Harper and Row New York, 415 pp

Hayes KF, Traina SJ (1998) Metal Ion speciation and its signification in ecosystem health. Soil chemistry and Ecosystem Health Spec Publ No 52, Soil Science Society of America, Madison, MI

Haygarth PM (1994) Global importance of global cycling of selenium. In: Frankenberger WT, Benson S (eds) Selenium in the Environment. Marcel Dekker New York, p 1–28

Heidweiller VML (1990) Fluoride removal methods. In: Frenken JE (ed) Proc Symp on Endemic Fluorosis in Developing Countries- Causes Effects and Possible Solutions. NIPG-TNO, p 51-85

Hewitt D, Neri LC (1980) Development of the "water story" some recent Canadian studies. J Environ Pathol Toxicol 4(2-3): 51-63

Heymann EW, Hartmann G (1991) Geophagy in moustached tamarins *saguinus mystax* (Platyrrhini: *Callitrichidae*) at the Rio Blanco. Peruvian Amazonia Primates 32: 533-537

Higgo JW, Haigh DG, Allen MR, Warwick P, Williams GM (1991) Iodine speciation and diffusion in a sand-groundwater system. Nuclear Sci Tech Tropical Report for the European communities EUR 13277, 33 pp

Hipkin J, Shaw PV (1999) Working with ores containing naturally occurring radioactive material. 3rd European ALARA Network Workshop Managing Internal Exposures

Hochella MF (1993) Surface chemistry structure and reactivity of hazardous mineral dust. In: Gutherie DG, Mossman B (eds) Reviews in Mineralogy 28: 275-308

Hodge HC, Smith FA, Gedalia I (1970) Excretion of fluorides In Fluorides and Human Health. WHO Geneva Monograph Series No 59: pp 158-159

Houghton JT (2001) Climate change 2001: the scientific basis. Contribution of Working Group I- 3rd assessment report of the Intergovernmental Panel on Climate Change, Cambridge University Press, Cambridge, New York

Huq SMI, Joardar JC, Parvin S, Correll R, Ravi Naid R (2006) Arsenic Contamination in food-chain: Transfer of arsenic into food materials through groundwater irrigation. J Health Popul Nutr 24(3): 305 - 316

Hutton LG, Lewis WJ (1980) Nitrate pollution of groundwater in Botswana. 6th WEDC Conf Water and Water and Engineering in Africa, 1-4

ICMR Task Force (1989) Epidemiological survey of endemic goitre and endemic cretinism. New Delhi Indian Council of Medical Research

IAEA (1996) International basic safety standards for protection against ionizing radiation and for safety of radiation sources. IAEA Safety Series 115, Vienna

IFA (International Fertilizer Industry Association) (2004) Fertilizer nutrient consumption. The world 1920/21 to 200/01 Outlook to 2030

IPCS (International Programme on Chemical Safety) (1987) Environmental Health Criteria, World Health Organization, Geneva, 236 pp.

Ishiga H, Dozen K, Yamazaki C, Ahmed F, Islam B, Rahan H, Satter A, Yamamoto H, Itoh K (2000) Geological constraints on arsenic contamination in Bangladesh. 5th Forum on Arsenic contamination. Yokohama Japan 2000

Islam MR, Lahermo P, Salminen R, Rojstaczer S, Peuraniemi V (2000) Lake and reservoirs water quality affected by metals leaching from tropical soils Bangladesh. Environ Geol 39 (10): 1083-1089

Iyengar GV, Ayengar ARG (1988) Human health and trace elements including effects on high-altitude populations. Ambio 17(1):31-35

Iyengar MAR (1990) The natural distribution of radium The environmental behaviour of radium. Technical Reports Series 310, International Atomic Energy Agency, Vienna, pp 59-128

Jaikrishan G, Andrews VJ, Thampi MV, Koya PK, Rajan VK, Chauhan PC (1999) Genetic monitoring of the human population from high level natural radiation areas of Kerala on the southwest coast of India-Prevalence of congenital malformations in new borns. Radiation Res 152 (6): 149-153

Jaworowski Z (2000) Ionizing radiation and radioactivity in the 20th century. Int Conf on radiation and its role in diagnosis and treatment. FICR-2000 Tehran Iran Oct 18-20

Jinadasa KBPN, Weerasooriya SVR, Dissanayake CB (1988) A rapid method for the defluoridation of fluoride-rich drinking at village level. Intern J Environ Studies 31: 305-312

Jinadasa KBPN, Dissanayake CB, Weerasooriya SVR (1991) Use of serpentinite in the defluoridation of fluoride-rich drinking water. Inter J Environ Studies 37: 43-63

Johns T (1986) Detoxification functions of geophagy and domestication of the potato. J Chem Ecology 12: 635-646

Johns T, Duquette M (1991) Detoxification and mineral supplementation as functions of geophagy. Am J Chin Nutr 53: 448-456

Johnson CC (2003a) Database of the iodine content of soils populated with data from published literature. (BGS commissioned report /CR/03/004N), 38 pp

Johnson CC, Fordyce FM, Stewart AG (2003b) Environmental Controls in iodine deficiency disorders. Project Summary (British Geological Survey-DFID) Commissioned report CR/03/058N

Johnson CC, Ge X, Green KA, Liu X (2000) Selenium distribution in the local environment of selected villages of the Keshan Disease belt Zhangjiakou District Hebei Province People's Republic of China. Applied Geochemistry 15: 385-401

Joiner MC (1994) Induced radio resistance: an overview and historical perspective. Int J Radiat Biol 65: 79-84

Juhasz, AL, Megharaj M, Naidu R (2000) Bioavailability: The major challenge (constraint) to bioremediation of organically contaminated soils. In: Remediation of hazardous waste contaminated soils, 2nd edition, volume 1: Engineering Considerations and Remediation Strategies, section 1-1: Engineering Issues in Waste Remediation, pp 217-241

Juhasz A, Smith E, Naidu R (2003) Estimation of human availability of arsenic in contaminated soils. In: Langley A Gilbey M Kennedy B (eds) Proc 5th National Workshop on the assessment of site contamination. National Environmental Protection Council Service Corporation, p 183-194

Kabata-Pendias A, Pendias H (1984) Trace elements in the soil and plants. CRC Press, Boca Raton, Florida USA

Kadurin S (1998) Minerals in human kidneys. Proceedings of the European Crystallographic Meeting, Praha, Czech Republic, August 16–20, 1998

Kajander EO, Çiftçioglu N (1998) Nanobacteria: an alternative mechanism for pathogenic intra- and extracellular calcification and stone formation. Proc Natl Acad Sci USA 95(14): 8274 - 8279

Kajubi SK (1971) Iodine in the Uganda environment. East Afr Med J 48: 427

Kang BT, Osiname DA (1985) Micronutrient problems in tropical Africa. Fertilizer Research 7: 131-150

Kapil U, Jayakumar PR, Singh P, Aneja B, Pathak P (2002) Assessment of iodine deficiency on Kottayam district Kerala State: a pilot study. Asia Pacific J Clin Nutr 11(1): 33-35

Karim MDM (2000) Arsenic in groundwater and health problems in Bangladesh. Water Research 34(1): 304-310

Karppanen H, Pennanen R, Passinen L (1978) Minerals coronary heart disease and sudden coronary death. Adv Cardiol 25: 9-24

Kartha CC, Valiathan MS, Eapen JT, Pathinam K, Kumary TV, Raman Kutty V (1993) Enhancement of cerium levels and associated myocardial lessons in hypomagnesaemic rats fed on cerium-adulterated diet. In: Valiathan MS Somers K Kartha CC (eds) Endomyocardial Fibrosis. Oxford University Press Delhi Bombay Calcutta Madras

Kau PMH, Smith DW, Binning P (1998) Experimental sorption of fluoride by kaolinite and bentonite. Geoderma 84: 89-108

Kau PMH, Smith, DW, Binning P (1997) Fluoride retention by kaolin clay. J Contaminant Hydrology 28: 267-288

Keay J (1993) Eating dirt in Venezuela. In: Keay J (ed) The Robinson Book of Exploration. Robinson London (taken from Alexander non Humboldt Personal Narrative of Travels to the Equinoctial Regions of America George Routledge and Sons London), p 342-350

Kehew AE (2000) Applied chemical hydrogeology. Prentice Hall, Upper Saddle River, New Jersey, USA, 365 pp

Keil U (1979) Hardness of drinking water (content of bulk and trace elements) and cardiovascular diseases (in German). Geographische Zeitschrift Sonderdruck. Geomedizin in Forschung und Lehre. Franz Steiner Verlag GmbH, Wiesbaden, p 59-76.

Kelly FC, Snedden WW (1958) Prevalence and geographical distribution of endemic goitre. WHO Bull 18: 5-173

Kelsey KT, Memisoglu A, Frenkel D, Liber HL (1991) Human lymphocytes exposed to low doses of X-rays are less susceptible to radiation-included mutagenesis. Mutat Res 263: 197-201

Kennedy TP, Dodson R, Rao NV, Bazer M, Tolley E, Hoidal JR (1989) Dust causing pneumoconiosis generate OH and produce haemolysis by acting as Fenton Catalysts. Archives of Biochemistry and Biophysics 269: 359- 364

Kerndorff H, Schnitzer M (1980) Sorption of metals on humic acid. Geochim Cosmochim Acta 44: 1701-1708

Kesavan PC (1996) Indian research on high levels of natural radiation: pertinent observations for further studies. In: L Wei T Sugahara Z Tao (eds) High levels of Natural Radiation. Radiation Dose and Health Effects: Beijing China, Elsevier Amsterdam, p 111-117

Ketch LA, Malloch D, Mahaney WC, Huffman MA (2001) Comparative microbial analysis and clay mineralogy of soils eaten by chimpanzees (*Pan Troglodytes schweinfurthii*) in Tanzania. Soil Biology and Biochemistry 33: 199-203

Kibblewhite ME (1982) The influence of trace element distribution and availability in soils on the occurrence of oesophageal cancer. In: Kibblewhite ME, Laker MC (eds) Trace element distribution in relation to oesophageal cancer in the Butterworth District Transkei. University of Fort Hare South Africa

Klaus G, Klaus-Hugi C, Schmid B (1998) Geophagy by large mammals at natural licks in the rain forest of Dzanga National Park Central African Republic. J Trop Ecol 14: 829-839

Kobayashi J (1957) On geographical relationship between the chemical nature of river water and death rate from apoplexy. Berichte des Ohara Instituts für landwirtschaftliche Biologie 11, Okayama University 14: 12-21

Kochipillai N, Ramalingaswamy V, Stanbury JB (1980) Southeast Asia. In: Stanbury JB, Hetzel BS (eds) Endemic goitre and endemic cretinism. John Wiley New Delhi, p 101-122

Köppen W (1936) Das geographische System der Klimate. Handbuch der Klimatologie Bd 1 Teil C

Koss V (1997) Umweltchemie. Springer, 288 pp

Kovalsky VV (1979) Geochemical ecology and problems of health. Phil Trans R Soc Lond B 288: 185-191

Kožíšek F (2003) Health significance of drinking water calcium and magnesium. National Institute of Public Health Czech Republic (http://wwwszucz/chzp/voda/pdf/hardness.pdf)

Kronberg BI, Fyfe WS, Leonardos OH Jr, Santos AM (1979) The chemistry of some Brazilian soils: Element mobility during intense weathering. Chem Geol 24:211-219

Kühnel RA (1987) The role of cationic and anionic scavengers in laterites. Chem Geol 60: 31-40

Kühnel RA, Roorda HJ, Steensma JJS (1978) Distribution and portioning of element in nickeliferous laterites. Bull BRGM II: 191-206

Kuklinsky B, Schweder R (1996) Acute pancreatitis - a free radical disease: reducing the lethality with sodium selenite and other antioxidants. J Nutr Environ Med 6: 393-394

Kumari PJ, Ashlesha P, Kodaate J, Vali SA (1995) Fluorine partition between coexciting soil and food stuff: A case study of selected fluorosis endemic villages of Kalpana Tehsil District, Chandrapur (MS). Gondwana Geol Hag 9: 21-27

Laftouhi N, Vanclooster M, Jalal M, Witam O, Aboufirassi M, Bahir M, Persoons E (2003) Groundwater nitrate pollution in the Essaouira Basin (Morocco). Comptes rendus Géoscience 335: 307-310

References

Låg J (ed) (1980) Geomedical aspects in present and future research. Universitetforlaget, Oslo Norway, 226 pp

Låg J (ed) (1983) Geomedical research in relation geochemical registration. Oslo Universitetforlaget, Oslo Norway

Laker MC (1979) Mineral element studies on soil and plant samples from low and high incidence districts. In: Van Rensburg SJ (ed) Environmental associations with oesophageal cancer in Transkei. (Tygerberg Medical Research Council South Africa)

Laker MC, Hensley MdeL, Beyers CP, Rensburg V (1980) Environmental associations with oesophageal cancer-an integrated model. S Afr Cancer Bulletin 24: 69-70

Lalonde JP (1976) Fluorine-an indicator of mineral deposits. Canadian Mining and Metallurgical Bulletin, May: 110-122

Landrum PF, Hayton WL, (1992) Synopsis of discussion session on the kinetics behind environmental bioavailability. In: Hamelink JL, Landrum PF, Bergman HL, Benson WH (eds) Bioavailability Physical Chemical and biological interactions. Proc. of a workshop held in Pellston Michigan, Lewis Publishers

Langer P, Greer MA (1977) Antithyroid substances and naturally occurring goitrogens. S Karger, Basel, New York

Latham MC, McGandy RB, McCann Mb, Stare FJ (1972) Fluoride. In: Scope manual on nutrition, 2nd edition, Kalamazoo, the Upjohn Co, p 61-63

Lauria DC, Godoy JM (2002) Abnormal high natural radium concentration in surface waters. J environ Radioact 61: 159-168

Lawrence AR (1986) Risk to groundwater supply from bit latrine Soakaways in Sri Lanka. British Geological Survey Hydrogeology Research Group. WD/05/86/21, Technical Report

Leutenegger F (1956) Changes in the ammonia and nitrate contents of tropical red loam as influenced by manuring and mulching during a period of one year. E African Agr J22: 81-87

Leon LA, Lopez AS, Vlek PLG (1985) Micronutrient problems in tropical Latin America. Fertilizer research 7: 95-129

Leonard RH (1961) Quantitative composition of kidney stones. Clinical Chemistry 7: 546-551

Licht OAB, Tarvainen T (1996) Multipurpose geochemical maps produced by integration of geochemical exploration data sets in the Parana Shield Brazil. Journal of Geochemical Exploration 56: 167-182

Linchenat A, Shirokova I (1964) Individual characteristics of the nickeliferous iron (laterite) deposits of the northeastern part of Cuba (Pinares de Mayari Nicaro Moa). Proceedings of the 22nd International Geological Congress 14: 169-187

Linsalata P, Eisenbud M, Franca EP (1986) Ingestion estimates of thorium and the light rare earth elements based on measurements of human faeces. Health Phys 50:163-167

Longvah T, Deosthale YG (1998) Iodine content of commonly consumed foods and water from the goitre-endemic North-East region of India. Food Chemistry 61(3): 327-333

Lovelock J (1988) The ages of Gaia. In: The Biography of our living earth. Oxford University Press

Lubkowska A, Zyluk B, Chlubek D (2002) Interactions between fluorine and aluminium. Fluoride 35(2): 73-77

Luo K, Ren D, Xu L, Dai S, Cao D, Feng F, Tan J (2003) Fluorine content and distribution pattern in Chinese coals. Int J of Coal Geology 57: 143-149

MacGregor LA (1998) The geochemistry of selenium in sedimentary environments examples from the UK and Jordan. Unpublished Ph D Thesis University of Reading UK

Mafa B (2003) Environmental hydrogeology of Francistown. Bundesanstalt für Geowissenschaften and Rohstoffe and Department of Geological Survey Botswana, 81 pp

Mahadeva K, Seneviratne DA, Jayatilake B, Shanmuganathan SS, Premachandra P, Nagaraja M (1968) Further studies on the problem of goitre in Ceylon. British J Nutr 22: 525-534

Mahaney WC (1993) Scanning electron microscopy of earth mined and eaten by mountain gorillas in the Virunga Mountains Rwanda. Primates 34: 311-319

Mahaney WC, Aufreiter S, Hancock RGV (1995) Mountain gorilla geophagy: A possible seasonal behaviour for dealing with effects of dietary changes. Int J Primatol 16: 475-488

Mahaney WC, Bezada M, Hancock RGV, Aufreiter S, Perez FL (1996) Geophagy of Holstein hybrid cattle in the Northern Andes Venezuela. Mount Res Dev 16: 177-180

Mahaney WC, Hancock RGV, Inoue M (1993) Geochemistry and clay mineralogy of soils eaten by Japanese Macaque. Primates 34: 85-91

Mahaney WC, Milner MW, Mulyano H, Hancock RGV, Aufreiter S, Reich M, Wink M (2000) Mineral and chemical analyses of soils eaten by humans in Indonesia. Int Journ of Environ Health Res 10(2): 93-109

Mahaney WC, Zippin J, Milner MW, Sanmugadas K, Hancock RGV, Aufreiter S, Campbell S, Huffman MA, Wink M, Mallock D, Kalm V (1999) Chemistry mineralogy and microbiology of termite mound soil eaten by chimpanzees of the Mahale Mountains Western Tanzania. J Tropical Ecology 15: 565-588

Mahesh DL (1993) Studies on iodine metabolism-role of dietary factors in endemic goitre biochemical mechanism of goitrogenecity. PhD thesis Osmania University Hyderabad India

Maier JAM (2003) Low magnesium and atherosclerosis: an evidence-based link. Molecular Aspects of Medicine 24: 137-146

Maithani PB, Gujar R, Banerjee R, Balaji BK, Ramachandran S, Singh R (1998) Anomalous fluoride in groundwater from the western part of Sirohi district Rajasthan and its crippling effects on human health. Current Science 74(9): 773-777

Mallick S, Rajagopal NR (1996) Groundwater development in the arsenic-affected alluvial belt of west Bengal-Some questions. Current Science 70: 956 -958

Malisa EP (2001) The behaviour of selenium in geological processes, Environmental Geochemistry and Health 23: 137–158

Mandal BK, Chowdhury TR, Samanta G, Mukherjee DP, Chanda CR, Saha K, Chakraborti D (1998) Impact of safe water for drinking and cooking on five arsenic-affected families for 2 years in West Bengal India. The Science of the Total Environ 218: 185-201

Manji F, Baelum V, Fejerskov O (1986) Dental fluorosis in an area of Kenya with 2 mg/l in the drinking water. J Dental Res 65: 654-662

Manning BA, Goldberg S (1997) Adsorption and stability of arsenic (III) at the clay mineral-water interface. Environ Sci Technol 31 (7): 2005-2011

Manten AA (1966) Historical foundations of chemical geology and geochemistry. Chemical Geology 1: 5-31

Mariappan P, Vasudevan T (2002) Domestic defluoridation techniques and sector approach for fluorosis mitigation. J Inst Publ Health Eng 1: 17-21

Marier JR (1978) Cardio-protective contribution of hard waters to magnesium intake. Rev Can Biol 37(2):115-125

Marshall SJ, Balooch M, Habelitz S, Balooch C, Gallagher R, Marshall GW (2003) The dentin-enamel junction- a natural multilevel interface. J European Ceramic Society 23: 2897-2904

Masironi R (1979) Geochemistry and cardiovascular diseases. Phil-Trans Roy Soc London Series B 288: 193-201

Mason B (1992) Victor Moritz Goldschmidt- Father of modern Geochemistry. Special Publication No 4: The Geochemical Society Special Publication, San Antonio, 184 pp

Matschullat J (2000) Arsenic in the geosphere- a review. Science of the Total Environment 249 (1-3): 297-312

Matsumoto H, Hirasawa E, Morimura S, Takahashi E (1976) Localization of aluminium in tea leaves. Plant Cell Physiol 17: 627-631

Mayland HF, James LF, Panter KE, Sonderegger JL (1989) Selenium in seleniferous environments. In: Jacobs LW (ed) Selenium in agriculture and the environment Madison WI USA: Soil Science Society of America Special Publication Number 23: 15–50

McArthur JM, Ravenscroft P, Safiullah S, Thirlwall MF (2001) Arsenic in groundwater testing pollution mechanisms for sedimentary aquifers in Bangladesh. Water Resources Research 37(1): 109-117

McCarrison R (1908) Observations on endemic cretinism on Chitral and Gilgit Valleys. Lancet 2: 1275-1280

McGreevy PD, Hawson LA, Habermann TC, Cattle SR (2001) Geophagia in horses: a short note on 13 cases. Appl Anim Behav Sci 71: 119-125

McNeal JM, Balistrieri LS (1989) Geochemistry and occurrence of selenium: an overview. In: Jacobs LW (ed) Selenium in agriculture and the environment Madison WI USA. Soil Science Society of America Special Publication No 23:1-14

Meharg AA, Rahman MDM (2003) Arsenic contamination of Bangladesh paddy field soils: implications for rice contribution to arsenic consumption. Environ Sci Technol 37: 229-234

Mills CF (1996) Geochemical aspects of the aetiology of trace element related diseases. In: Appleton JD, Fuge R, McCall GJH (eds) Environmental Geochemistry and Health. Geological Society Special Publication UK,113, p 1-5

Miller EC, Miller JA (1972) Approaches to the mechanism and control of chemical carcinogenesis. In: Environment and Cancer. 24th Annual Symposium on Fundamental Cancer Research. Williams and Wilkins Baltimore Maryland, pp 5–39

Ministry of Health (PCR) (1997) Annual report of endemic diseases prevention in China. Beijing (in Chinese)

Minoguchi G (1974) The correlation of chronic toxic effects in tropical and subtropical areas between fluoride concentration in drinking water climate especially mean annual temperature. Proc Symp on Fluorosis (Hyderabad, India), 175-186

Miyake Y, Iki M (2004) Lack of association between water hardness and coronary heart disease mortality in Japan. Int J Cardiol 96: 25-28

Moffat AS (1990) China a living lab for epidemiology. Science 248: 553-555

Mohanty AK, Sengupta D, Das SK, Saha SK, Vans KV (2004a) Natural radioactivity and radiation exposure in the high background area at Chhatrapur beach placer deposit of Orissa India. Journal of Environmental Radioactivity 75: 15-33

Mohanty AK, Sengupta D, Das SK, Vijayan V, Saha SK (2004b) Natural radioactivity in the newly discovered high background radiation area on the eastern coast of Orissa India. Radiation Measurements 38: 156-165

Mortazavi SMJ (2003) Lessons learned from Ramsar Studies. Karlsruhe University Dec 8 2003 Germany

Mortazavi SMJ, Ghiassi Nejad M, Beitollahi M (2001) Very high background radiation areas (VHBRAs) of Ramsar: Do we need any regulations to protect the inhabitants? Proc 34th mid-year meeting Radiation Safety and ALARA-Considerations for the 21st Century California USA, 177-182

Munasinghe T, Dissanayake CB (1982) A plate tectonic model for the geologic evolution of Sri Lanka. J Geol Soc India 28: 369-380

Munoz JL, Ludington SD (1974) Fluoride-hydroxyl exchange in biotite. American Journal of Science 274: 396-413

Muramatsu Y, Ohmoto Y (1986) Iodine-129 and iodine-127 in environmental samples collected from Tokaimura/Ibaraki, Japan. Sci Total Environ 48: 33-43

Muramatsu Y, Sumiya M, Ohmoto Y (1983) Stable iodine contents in human milk related to dietary algae consumption. Hoken Butsuri 18: 113-117

Muramatsu Y, Wedepohl KH (1998) The distribution of iodine in the earth's crust. Chemical Geology 147: 201-216

Muramatsu Y, Yoshida S, Fehn U, Amachi S, Ohmomo Y (2004) Studies with natural and anthropogenic iodine isotopes: distribution and cycling of iodine in the global environment. J Environ Radioactivity 74: 221-232

Muramatsu Y, Yoshida S (1999) Effects of microorganisms on the fate of iodine in the soil environment. Geomicrobiol J 16: 85-93

Nair KR, Manji F, Gitonga JN (1984) The occurrence and distribution of fluoride in groundwaters of Kenya. East Afr Med J 61: 503-512

Nair MK, Nambi KSV, Amma NS, Gangadharn P, Jayalekshmi P, Jayadevan S, Cherian V, Reghuram KN (1999) Population study in the high natural background radiation area in Kerala India. Radiation Research 152: 145–148

Nanayakkara D, Chandrasekara M, Wimalasiri WR (1999) Dental fluorosis and caries incidence in rural children in a high fluoride area in the dry zone of Sri Lanka. Ceylon Med J 42: 13-17

National Academy of Science (1974) Geochemistry and the environment. Vol I Washington DC

National Academy of Science (1978) Geochemistry and the environment. Vol II Washington DC

National Academy of Science (1977) Drinking water and health, Assembly of Life Science. Nat Acad Sci, Washington DC

National Research Council (1978) Nitrates: an environmental assessment. National Academic Press, Washington DC

Nawlakhe WG, Bulusu KR (1989) Nalgonda technique- a process for removal of excess fluoride from water. Water Quality Bull 14: 218-220

NEFESC (2000) Guide for incorporating bioavailability adjustments into human health and ecological risk assessments at US Navy and Marine corps facilities Part 1: Metals bioavailability Naval Facilities, NEFESC Engineering Command Washington DC, UG 2041-Env, 56 pp

Neri LC, Hewitt D, Schreiber GB, Anderson TW, Mandel JS, Zdrojewsky A (1975) Health aspects of hard and soft waters. J Amer Water Works Assoc 67: 403-409

Neri LC, Johansen HL, Talbot FDF (1977) Chemical content of Canadian drinking water related to cardiovascular health. Publication of Univ Ottawa Canada

Nève J (1996) Selenium as a risk factor for cardiovascular diseases. J Cardiovascular Risk 3: 42-47

Ngo DB, Dikassa L, Okitolonda W, Kashala TD, Gervy C, Dumont J, Vanovervelt N, Contempre B, Diplock AT, Peach S, Vanderpas J (1997) Selenium status in pregnant women of a rural population (Zaire) in relationship to iodine deficiency. Tropical Medicine and International Health 2(6): 572-581

Nickson R, McArthur J, Burgess W, Ahmed KM, Ravescroft P, Rahman M (1998) Arsenic poisoning of Bangladesh groundwater. Nature 359: 338

Nickson RT, McArthur JM, Ravenscroft P, Burgess WB, Ahmed KM (2000) Mechanism of arsenic poisoning of groundwater in Bangladesh and West Bengal. Applied Geochemistry 15:403-413

Nkotagu H (1996) Origins of high nitrate in groundwater in Tanzania. J African Earth Science 21(4): 471-478

Nopakun J, Messer HH (1990) Mechanism of fluoride absorption from the rat small intestine. Nutr Res 10: 771-780

Norboo T, Angchuk PT, Yahya M, Kamat SR, Pooley FD, Corrin B, Kerr IH, Bruce N, Ball KP (1991) Silicosis in a Himalayan village population role of environmental dust. Thorax 46: 341-343

Nordstrom DK (2002) Worldwide occurrences of arsenic in groundwater. Science 296: 2143

Nriagu JO, Pacyna JM (1988) Quantitative assessment of worldwide contamination of air water and soils by trace metals. Nature 333: 134-139

Nunn JH, Regg-Gunn AJ, Ekanayake L, Saparamadu KDG (1994) Prevalence of developmental effects of enamel in areas with differing water fluoride levels and socio-economic groups in Sri Lanka and England. Int Dent J 44: 165-173

Oh CK, Lücker PW, Wetzelsberger N, Kuhlmann P (1986) The determination of magnesium calcium sodium and potassium in assorted foods with special attention to the loss of electrolytes after various forms of food preparations. Magnesium Bulletin 8: 297-302

Olivieri G, Bodycote J, Wolff S (1984) Adaptive response of human lymphocytes to low concentrations of radioactive thymidine. Science 223: 594-597

Ophang RH (1990) Fluoride. In: Brown ML (ed) Present knowledge in nutrition. 6th edition, Washington DC ILSI Press. p 274-278

O'Reilly SE, Strawn DG, Sparks DL (2001) Residence time effects on arsenate adsorption/desorption mechanisms on goethite. Soil Science Society of America Journal 65(1): 67-77

Oremland, RS (1994) Biogeochemical transformations of selenium in anoxic environments. In: Selenium in the Environment, Frankenberger JR, Benson S (eds) Marcel Dekker, NY, p 389-420

Oremland RS, Stolz JF (2003) The ecology of arsenic. Science 300: 939-944

Oremland RS, Dowdle PR, Hoeft S, Sharp JO, Schaefer JK, Miller LG, Swizerblum J, Smith RL, Bloom NS, Wallschlaeger D (2000) Bacterial dissimilatory reduction of arsenate and sulphate is meromictic Mono Lake California. Geochim Cosmochim Cosmochimica Acta 64: 3073-3084

Oremland RS, Hollibaugh JT, Maest AS, Presser TS, Miller LG, Culbertson CW (1989) Selenate reduction to elemental selenium by anaerobic bacteria in sediments and culture: Biogeochemical significance of a novel sulphate-independent respiration. Appl Environ Microbiol 55: 2333-2343

Oremland RS, Newman DF, Wail BW, Stolz JF (2002) Bacterial respiration of arsenate and its significance in the environment. In: Frankenberger Jr WT (ed) Environmental Chemistry of Arsenic. Marcel Dekker NY, p 273-296

Osiname O, van Gijn H, Vlek PLG (1983) Effect of nitrification inhibitors on the fate on and efficiency of nitrogenous fertilizers under simulated humid tropical conditions. Trop Agric (Trinidad) 60: 211-217

Ozha DD, Varshney CP, Bohra JL (1993) Nitrate in groundwaters of some districts of Rajasthan. Indian J Environ Health 35: 15-19

Padmasiri JP, Dissanayake CB (1995) A simple defluoridator for removing excess fluorides from fluoride-rich drinking water. Int Journ Environ Health Research 5: 153-160

Pal T, Mukherjee PK, Sengupta S, Bhattacharyya AK, Shome S (2002) Arsenic pollution in groundwater of West-Bengal India; An insight into the problem by subsurface analysis. Gondwana Research 5: 501-512

Pandav CS, Kochupillai N (1982) Endemic goitre in india: prevalence etiology Attendant disabilities and control measures. Indian J Pediat 50: 259-271

Pandav CS, Rasheed M, Solih I, Saeed M, Shaheed M, Awal A, Anand K, Shreshta R (1999) Iodine deficiency disorders in the Maldives: A public health problem. Asia Pacific Journal of Clinical Nutrition 8(1): 9-12

Paschoa AS, Godoy JM (2002) The areas of high natural radioactivity and TENORM wastes. Int Cong Series 1225: 3-8

Paul AC, Pillai PMB, Haridasan PP, Radhakrishnan S, Krishnamony S (1998) Population exposure to airborne thorium at the high natural radiation areas in India. J Environ Radioactivity 40(3): 251-259

Pecora WT, Hobbs SW, Murata KJ (1949) Variations in garnierite from the nickel deposit near Riddle Oregon. Econ Geol 44: 13-23

Pedro G (1985) Grandes tendencies des sols mondiaux. Cultivar 184:78-81

Peijnenburg WJGM, Posthuma L, Eijsackers HJP, Allen HE (1997) A conceptual framework for implementation of bioavailability of metals for environmental management purposes. Ecotoxicology and Environmental Safety 37: 163-172

Pendrys DG (2001) Fluoride ingestion and oral health. Nutrition 17: 979-980

Peterson S, Légué F, Tylleskär T, Kpizingui E, Rosling H (1995) iodine deficiency disorders in the Central African Republic. Nutrition Research 15: 803-812

Phantumvanit P, Songpaisan Y, Moller IJ (1998) A defluoridator for individual households. World Health Forum 9: 555-558

Pharoah POD (1985) The epidemiology of endemic cretinism. In: Follett BK, Ishi S, Chandola A (eds) The Endocrine System and the Environment. Springer, Berlin, p 315–322

Pickering WF (1985) The mobility of soluble fluoride in soils. Environ pollut Ser B Chem Phys 9: 281-308

Piyasena RD (1979) Endemic goitre in Sri Lanka. In: Proc current thyroid problems in southeast Asia and Oceania. Asia and Oceania Thyroid Assoc 1: 30-34

Plant JA, Baldock JW, Smith B (1996) The role of geochemistry in environmental and epidemiological studies in developing countries: a review. In: Appleton JD, Fuge R, McCall GJH (eds) Environmental Geochemistry and health. Geological society Special Publication UK, 113: p 7-22

Plant J, Smith D, Smith B, Williams L (2000) Environmental geochemistry at the global scale. J Geol Soc Lond 157: 837-849

Plant J, Thornton I (1986) Geochemistry and health in the United Kingdom. In: Thornton I (ed) Proc First Int Symp on Geochemistry and Health Science Reviews, Northwood, p 5-15

Pocock SJ, Shaper AG, Cook DG, Packham RF, Lacey RF, Powell P, Russell PF (1980) British regional heart study: geographic variations in cardiovascular mortality and the role of water quality. Br Med J 250: 1243-1249

Pollycove M, Feinendegen LE (2001) Biological responses to low doses of ionizing radiation: detriment versus hormesis. J Nucl Med 42(9): 26-37

Polomski J, Flühler H, Blaser P (1982) Accumulation of airborne fluoride in soils. J Environ Qual 11: 457-461

Powell JJ, Ainley CC, Harvey RSJ, Mason JM, Kendall MD, Sankey EA, Dhillon AP, Thompson RPH (1996) Characterization of inorganic microparticles in pigments cells of human gut associated lymphoid tissue. Gut 38: 390-395

Price EW (1988) Non-filarial elephantiasis-confirmed as a geochemical disease and renamed podoconiosis. Ethiopian Medical Journal 26: 151-153

Price EW (1990) Podoconiosis non-filarial elephantiasis. Oxford University Press, 144 pp

Price EW, Bailey D (1984) Environmental factors in the aetiology of endemic elephantiasis of the lower legs in tropical Africa. Tropical Geogr Med 36: 1-5

Price EW, Henderson WJ (1978) The elemental content of lymphatic tissues of barefooted people in Ethiopia with reference to endemic elephantiasis of their lower legs. Trans Roy Soc Trop Med Hyg 72: 132-136

Price EW, Henderson WJ (1981) Endemic elephantiasis of the lower legs in the United Cameroon Republic. Tropical and Geographical Medicine 33: 23-29

Price EW, Plant DA (1990) The significance of particle size of soils as a risk factor in the etiology of podoconiosis. Trans Roy Soc Trop Med Hyg 84: 885-886

Pulgar T, Rai BM, Shankar R, Rai BM (1992) Iodine deficiency disorders in Sikkim: current status and future strategy. Thyroid Centre Government General Hospital Namchi, South Sikkim, 67 pp

Pungrassami T (1970) Preliminary geochemical study of nickel in soil over serpentinite at Ban Ragam Chanthaburi Province, Thailand. Seminar, 210-214

Raghava Rao KV, Bogoda KR, Nilaweera NS, de Silva PHDS (1987) Mapping high fluoride contents in groundwater of Sri Lanka. National Water Supply and Drainage Board Sri Lanka and WHO (Report)

Ralston SL (1986) Feeding behaviour. Vet Clin North Am Equine Pract 2(3): 609-621

Ramesam V, Rajagopalan K (1985) Fluoride ingestion into the natural waters of hard rock areas Peninsular India. J Geol Soc India 26: 125-132

Rao EVSP, Puttanna K (2000) Nitrates agriculture and environment. Current Science 79: 1163-1168

Rao KV, Khandekar AK, Vaidyanadham D (1973) Uptake of fluoride by water hyacinth *Eichhornia crissipes*. Indan J Exp Biol 11: 68-69

Rashid MA, King LH (1970) Major oxygen containing functional groups present in humic and fulvic acid fractions isolated from contrasting marine environments. Chemical Geology 34: 193-201

Ravenscroft P, McArthur JM, Hoque B (2001) Geochemical and palaeohydrological controls on pollution of groundwater by arsenic. In: Chappell WR, Abernathy CO, Calderon R (eds) Arsenic Exposure and Health Effects IV, Elsevier Science Ltd, p 53-77

Rayman MP (2000) The importance of selenium to human health. Lancet 356: 233-241

Razaq IB, Fahad AA, Al-Hadeethi AA, Tawseek HA (1987) The role of micro organisms in radioiodine retention by Iraqi calcareous soils. In: Proc 9th Int Symp on soil Biology Budapest, pp 793-804

Rea RE (1979) A rapid method for the determination of fluoride in sewage sludges. Water Pollution Control 78: 139-142

Reddy DR (1985) Some observation on fluoride toxicity. Nimhans J 3: 79-86

Reddy DB, Rao CM, and Sarada D (1969) Endemic fluorosis. J Indian Med Assoc 53: 275-281

Reedman AJ, Calow RC, Mortimer C (1996) Geological surveys in developing countries- Strategies for assistance. Brit Geological Survey Technical Report WC/96/20

Reilly C (1996) Selenium in food and health. Blackie Academic and Professional London

Reilly C, Henry J (2000) Geophagia: why do humans consume soil? Nutrition Bulletin 25: 141-144

Reimann C, Bjorvatn K, Frengstad B, Melaku Z, Tekle-Haimanot R, Siewers U, (2003) Drinking water quality in the Ethiopian section of the East African Rift Valley I-data and health aspects. Sci Total Environ 311: 65-80

Rigand O, Papadopoulo D, Moustacchi E (1993) Decreased deletion mutation in radioadapted human lymphoblasts. Radiat Res 133: 94-101

Robertson HE, Riddell WA (1949) Cyanosis of infants produced by high nitrate concentration in rural waters of Saskatchewan. Can J Public Health 40: 72-77

Rose EA, Porcerelli JH, Neale AV (2000) Pica: Common but commonly missed. Journal of the American Board of Family Practice 13: 353-358

Rotruer JT, Poue AL, Ganther HE, Swanson AB, Hafeman DG, Hoeskstra WG (1993) Selenium: Biochemical role as a component of glutathione peroxidase. Science 179: 558–90

Royal Society of London (1983) Environmental geochemistry and health. A report in summary Royal Soc London

Rozier RG (1994) Epidemiologic indices for measuring the clinical manifestation of dental fluorosis: overview and critique. Adv in Dental Research 8: 39-55

Rubenowitz E, Molin I, Axelsson G, Rylander R (2000) Magnesium in drinking water in relation to morbidity and mortality from acute myocardial infarction. Epidemiology 11(4): 416-421

Rylander R (1996) Environmental magnesium deficiency as a cardiovascular risk factor, J. Cardiovasc. Risk 3: 4-10

Safiullah S (1998) CIDA Arsenic project report: Monitoring and mitigation of arsenic in the groundwater of Faridpur Municipality. Jahangirnagar University Dhaka, Bangladesh, 96 pp

Saha D, Sharma CB (2002) Evidence of rise in fluoride concentration in ground water with time in marginal Alluvial area of mid- Ganga basin. In: Rao BV et al. (eds) Proc Intl Conf on Hydrology and Watershed Management 18-20 Dec Hyderabad, p 609-618

Saha JC, Dikshit AK, Bandyopadhyay M, Saha KC (1999) A review of arsenic poisoning and its effect on human health. Critical Rev Environ Sci Technol 29(3): 281-313

Saha KC (2003) Review of arsenicosis in West Bengal India-a clinical perspective. Critical Rev Environ Sci Technol 30(3): 127-163

Saikat SQ, Selim AM, Kessi J, Wehrli E, Hanselmann KW (2001) Transformation of arsenic compounds by bacteria from groundwater sediments of Bangladesh. (www.unizh.ch/~microeco/uni/kurs/mikoek/results/project2/arsen.html)

Saiyed HN, Sharma YK, Sadhu HG, Norboo T, Patel PD, Patel TS, Venkaiah K, Kashyap SK (1991) Non-occupational pneumoconiosis at high altitude villages in central Ladakh. Br J Industrial Med 48: 825-829

Sanderson BJ, Morley AA (1986) Exposure of human lymphocytes to ionizing radiation reduces mutagenesis by subsequent ionizing radiation. Mutat Res 164: 347-351

Savenko AV (2001) Interaction between clay minerals and fluoride-containing solutions. Water Resources 28: 274-277

Schindler DW, Mills KH, Malley DF, Findlay DL, Shearer JA, Davies IJ, Turner MA, Linsey GA, Cruikshank DR (1985) Long-term ecosystem stress-the effects of years of experimental acidification on a small Lake. Science 228: 1395-1401

Schnetger B, Muramatsu Y (1996) Determinations of halogens with special reference to iodine in geological and biological samples using pyrohydrolysis for preparation and inductively coupled plasma mass spectrometry and ion chromatography for measurement. Analyst 121: 1627-1631

Schnitzer M (1978) Humic substances: chemistry and reactions. In: Schnitzer M, Khan SU (eds) Soil Organic Matter. Elsevier Amsterdam, p 1-64

Schrauzer GN, White DA, Schneider CJ (1977) Cancer mortality correlation studies III-Statistical association with dietary selenium intakes. Bioinorganic chemistry 7: 35-56

Schroeder HA (1960) Relation between mortality from cardiovascular disease and treated water supplies: variations in states and 163 largest municipalities of the United States. J American Med Assoc 172: 1902-1908

Schuiling RD (1998) Geochemical engineering; taking stock. J Geochemical Exploration 62: 1-28

Schwartz BF, Schenkman NS, Bruce JE, Leslie SW, Stephen WL, Stroller ML (2002) Calcium nephrolithiasis: Effect of water hardness on urinary electrolytes. Urology 60(1): 23-27

Schwarz K, Foltz CM (1957) Selenium as an integral part of factor 3 against dietary necrotic liver degeneration. J Am Chem Soc 79: 3292-3293

Selinus O (1988) Biogeochemical mapping of Sweden for geomedical and environmental research. In: Thornton I (ed) Geochemistry and Health Science. Reviews Northwood Ltd, p 13-19

Setz EZF, Enzweiler J, Solferini VN, Amêndola MP, Berton RS (1999) Geophagy in the golden-faced saki monkey (*Pithecia pithecia chrysocephala*) in the Central Amazon. J Zool Lond 247: 91-103

Shacklette HT, Boerngen JG (1984) Element concentrations in soils and other surficial materials of the conterminous United States. USGS Professional Paper 1270, 105 p

Shaohua W, De Long R (2000) The final report of the expanded iodine dripping project in Xinjiang Uygur Autonomous Region 1997-1999. Health Bureau of Xinjian Uygur Autonomous Region China

Sharma CB (2001) Fluoride and fluorosis in Bihar and Jharkhand. National Seminar on Environment and Water Resources Management Bihar State Productivity Council Patna, p 30-34

Sharma SK (2003) High fluoride in groundwater cripples life in parts of India. Diffuse Pollution Conference Dublin 2003, Section 7, pp 51-52

Sheppard MI, Hawkins JL (1995) Iodine and microbial interactions in an organic. soil J Environ Radioactivity 29: 91-109

Shinonaga T, Gerzabek MH, Strebl F, Muramatsu Y (2001) Transfer of iodine from soil to cereal grains in agricultural areas of Austria. Sci Total Environ 267: 33-40

Shuster J, Finlayson B, Scheaffer R, Sierakowski R, Zoltek J, Dzegede S (1982) Water hardness and urinary stone disease. J Urol. 128(2):422–425

Siddqui AH (1955) Fluorosis in Nalgonda district Hyderabad-Deccan. British Med J 2: 1408-1413

Simulian JC, Motiwala S, Sigman RK (1995) Pica in a rural obstetric population. South Med J 88: 1236-1240

Singh B, Singh Y, Sekhon GS (1995) Fertilizer-N use efficiency and nitrate pollution of groundwater in developing countries. J Contam Hydro 20: 167–184

Singh PP, Barjatiya MK, Dhing S, Bhatnagar DS, Kothari R, Dhar S (2001) Evidence suggesting that high intake of fluoride provokes nephrolithiasis in tribal populations. Urological Research 29(4): 238-244

Smedley PL, Kinniburgh DG (2002) A review of the source behaviour and distribution of arsenic in natural waters. Applied Geochemistry 17: 517-568

Smedley PL, Edmunds WM, West JM, Gardner SJ, Pelig-Ba KB (1995) 2 Health problems related to groundwaters in the Obuasi and Bolgatanga areas Ghana. British Geological Survey Technical Report, WC/95/43, 122 pp

Smil V (2000) Cycles of life- Civilization and the biosphere (revised edition) Scientific American Library, New York

Smith B (1998) Cerium and endomyocardial fibrosis in tropical terrain: project summary report. British Geol Surv Report, WC/98/026

Smith B, Chenery SRN, Cook JM, Styles MT, Tiberindwa JV, Hampton C, Freers J, Rutakinggirwa M, Sserunjogi L, Tomkins A, Brown CJ (1998) Geochemical and environmental factors controlling exposure to cerium and magnesium in Uganda. J Geochem Expl 65: 1-15

Snyder GT, Fehn U (2002) Origin of iodine in volcanic fluids-^{129}I results from the central America Volcanic Arc. Geochimica et Cosmochimica Acta 66(21): 3827-3828

Sohrabi M (1996) World high level natural radiation and/or radon prone areas with special regards to dwellings. In: Wei L, Suahara T, Tao Z (eds) Proc 4th International Conference on High Levels of Natural Radiation (ICHLNR), Beijing, China, pp 3-7

Sokol E, Nigmatuline E, Maksimova N, Chiglntsev A (2005) $CaC_2O_4·H_2O$ spherulites in human kidney stones. European Journal of Mineralogy 17: 285-295

Sompura K (1998) Study on prevalence and severity of chronic fluoride intoxication in relation to certain determinants of fluorosis. PhD thesis ML Sukhadia University Udaipur (Rajasthan), India

Spencer H, Osis D, Wiatrowski E (1975) Retention of fluoride with time in man. Clin Chem 21: 613-618

Sprent JI (1987) The ecology of the nitrogen cycle. Cambridge Studies in Ecology, Cambridge University Press 160 pp

Stanbury JB, Hetzel BS (1980) Endemic goitre and endemic cretinism: Iodine Nutrition in Health and Disease. John Wiley New York

Stanton MF, Layard M, Tegeris A, Miller E, May M, Morgan E, Smith A (1981) Relation of particle dimension to carcinogenicity in amphibole asbestos and other fibrous minerals. Journal of the National Cancer Institute 67: 965-975

Stevens DP, McLaughlin MJ, Alston AM (1997) Phytotoxicity of aluminium-fluoride complexes and their uptake from solution culture by *Avena sativa* and *Lycopersicon esculentum*. Plant and Soil 192: 81-93

Stewart AG (1990) For debate: Drifting continents and endemic goitre in northern Pakistan. British Medical Journal 300: 1507-1512

Stewart AG, Pharoah POD (1996) Clinical end epidemiological correlates of iodine deficiency disorders. In: Appleton JD, Fuge R, Mc Call GJH (eds) Environmental Geochemistry and Health. Geol Soc Sp Publ UK,113, p 201-211

Stewart AG, Carter J, Parker A, Alloway BJ (2003) The illusion of environmental iodine deficiency. Environmental Geochemistry and Health 25: 165-170

Stormer JC, Carmichael ISE (1971) Fluorine-hydroxyl exchange in apatite and biotite: A potential igneous geothermometer. Contr Min and Pet 31: 159-167

Streit B (1992) Bioaccumulation processes in ecosystems. Cellular and Molecular Life Sciences 48: 955-970

Streit B, Stumm W (1993) Chemical properties of metal and the process of bioaccumulation in terrestrial plants. In: Markert B (ed) Plants as Biomonitors for Heavy Metal Pollution of the Terrestrial Environment. Verlag Chemie, Weinheim, New York

Strunz M (1970) Mineralogische Tabellen Akademische Verlagsgesellschaft. Geest und Portig KG, Leipzig

Stumm W, Morgan JJ (1996) Aquatic Chemistry. John Wiley and Sons New York

Sulton RN (1988) Fluoridation: a fifty-year old accepted but unconfirmed hypothesis. Med Hypotheses 27: 153-156

Sunde RA (1997) Selenium. In: O'Dell BL, Sunde RA (eds) Handbook of Nutritionally Essential Mineral Elements. Marcel Dekker, New York, p 493–556

Susheela AK (1998) Fluorosis management programme in India. (Presentation made in British Parliament to the all party group against fluoridation), October 1998

Susheela AK (2003) Treatise on fluorosis. 2nd edition, Fluorosis Research & Rural Development Foundation, Delhi, India

Tan J (1989) The Atlas of endemic diseases and their environments in the People's Republic of China. Science Press Beijing

Tan J, Zhu W, Li R (1989) Chemical endemic diseases and their impact on population in China. Environmental Sciences China 11: 107–114

Tan J, Zhu W, Wang W, Li R, Hou S, Wang D, Yang L (2002) Selenium in soil and endemic diseases in China. The Science of the Total Environment 284: 227-235

Tandia AA, Diop ES, Gaye CB (1999) Nitrate groundwater pollution in suburban areas; example of groundwater from Yeumbeul Senegal. J African Earth Science 29(4): 809-822

Tannenbaum SR, Correa P (1985) Nitrate and gastric cancer risks. Nature 317: 675-676

Tanskanen H, Gustavsson N, Koljonen T, Noras P (1988) The geochemical atlas as a means to establishing the balance of nutritional elements in Finnish soils. In: Thornton I (ed) Geochemistry and health: Proc of the 2nd International Symposium. London UK 22-24 April 1987, Northwood Science Reviews Ltd, p 1-11

Tateo F, Summa V, Bonelli CG, Bentivenga G (2001) Mineralogy and geochemistry of herbalist's clays for internal use simulation of the digestive process. Applied Clay Science 20: 97-109

Teitge JE (1990) Incidence in myocardial infarct and mineral content of the drinking water. Z Gesmte Inn Med 45: 478-485

Ten Cate JM, Featherstone JDB (1996) Physicochemical aspects enamel interactions. In: Fejerskov Ø, Ekstrand J, Burt BA (eds) Fluoride in Dentistry (2nd ed). Copenhagen Munksgaard, p 252-272

Teotia SPS, Teotia M (1992) Endemic fluoride bones and teeth uptake fluorosis in India. Manuscript Report -7, Institute of Social Sci, p 52-61

Thilly CH (1992) Goitre. In: Janssens PG, Kivits M, Vuylsteke J (eds) Médecine et Hygiène en Afrique Centrale de 1885 à nos jours. Fondation Roi Baudouin. Masson Paris Vol II, p 663-674

Thompson JM, Scott ML (1969) Role of selenium in the nutrition of the chick. J Nutr 97: 335-342

Thornton I (1987) Mapping of trace elements in relation to human disease. Clin Nutr 6(3): 97-104

Thornton I (1993) Environmental geochemistry and health in the 1990s: a global perspective. Applied Geochemistry Suppl 2: 203-210

Tiwari BK, Ray I, Malhotra RL (1998) Policy guidelines on natural Iodine Deficiency Disorders control programme. Nutrition and IDD cell New Delhi Directorate of Health Service, Ministry of Health and Family Welfare Government of India, p 1-22

Torriani A (1990) From cell membrane to nucleotides the phosphate regulon in *Escherichia coli*. Bioessays 12: 371-376

Tóth J (1963) A theoretical analysis of groundwater flow in small drainage basins. Journal of Geophysical Research 68 4795-4812

Tóth J (1999) Groundwater as a geologic agent: An overview of the causes processes and manifestations. Hydrogeology Journal 7:1-14

Townsend M, Rule AC, Meyer MA, Dockstader CJ (2007) Teaching the nitrogen cycle and human health interactions. Journal of Geoscience Education, March, 2007

Trescases JJ (1992) Chemical weathering. In: Butt CRM, Zeegers H (eds) Regolith exploration geochemistry in tropical and subtropical terrains. Handbook of Exploration Geochemistry: Volume 4, Elsevier Amsterdam, p 25-40

Tulpule PG (1969) Iodine and fluorine in drinking water. Indian J Nutr Dietet 6:229-233

Turner CH, Akhter MP, Heaney RP (1992) The effect of fluoridated water on bone strength. J Orthop Res 10(4): 581-587

Turekian KK, Wedepohl KH (1961) Distribution of the elements in some major units of the Earth's crust. Geol Soc Amer Bull 72: 175-192

Ukaonu C, Hill DA, Christensen F (2003) Severe hypokalemia in pregnancy due to clay ingestion. Obstet Gynecol 102:1169-117

Uma KO (1993) Nitrates in shallow (regolith) aquifers around Sokoto Town Nigeria. Environ Geol Water Sci 21: 70-76

Underwood EJ (1962) Trace elements in human and animal nutrition. Academic Press New York London

Underwood EJ (1977) Trace elements in Human and Animal Nutrition. Academic Press London

UNSCEAR (1982) Ionizing radiation source effects and biological effects. Report to the General Assembly with annexes 107-140, United Nations New York

UNSCEAR (1988) Source and effects of ionizing radiation. United Nations Scientific Committee on the Effect of Atomic Radiation, United Nations New York

UNSCEAR (1993) Source and effects of ionizing radiation. United Nations Scientific Committee on the Effect of Atomic Radiation, United Nations New York

UNSCEAR (2000) Source and effects of ionizing radiation. United Nations Scientific Committee on the effect of atomic radiation, United Nations New York

van Breemen N, Burrough PA, Velthorst EJ, van Dobben HF, De Wit T, Ridder TB, Reijnders HFR (1982) Soil acidification from atmospheric ammonium sulphate in forest canopy through fall. Nature 299: 548–550

van der Hoek W, Ekanayake L, Rajasooriya L, Karunaratne R (2003) Source of drinking water and other risk factors for dental fluorosis in Sri Lanka. Int J Environ Health Research 13: 285-293

Vanderpas JB, Contempre B, Duale NL et al (1990) Iodine and selenium deficiency associated with cretinism in northern Zaire. Am J Clin Nutr 52: 1987-1993

Vannappa B, Govindaiah S, Kariyanna H (1999) Fluoride problem in drinking water of Karnataka: an overall appraisal and remedial measures. In: Gyani KC, Vaish AK, Vaish P (eds) Proc National Seminar Environment and Health, p 17-22

Valiathan MS, Kartha CC, Nair RR, Shivakumar K, Eapen JT (1993) Geochemical basis of tropical endomyocardial fibrosis. In: Valiathan MS, Somers K, Kartha CC (eds) Endomyocardial Fibrosis Oxford England: Oxford University Press, p 98-110

Vijay Kumar V, Sai CST, Swami MSR, Rao PLKM (1993) Nitrates in groundwater sources in Medchal Block of Andhra Pradesh. Indian Journal of Environ Health 35: 40-46

Villa A, Carrasco G, Valenzuela A, Garrido A (1992) The effects of calcium on disoduim monofluorophosphate absorption from the gastrointestinal tracts of rats. Res Commun Chem Pathol Pharmacol 77: 367-374

Vinogradov AP (1938) Biogeochemical provinces and endemics. Doklady Akademii Nauk SSSR 18: 4-5 (in Russian)

Vinogradov AP, Lapp MA (1971) Use of iodine haloes to search for concealed mineralization. Vestnik-Leningradskii Universitet Seriia Geologii i Geografii No 24: 70-76 (in Russian)

Vitanage PW (1989) Precambrian and late lineament tectonic events in Sri Lanka. In: Recent Advances in Precambrian Geology in Sri Lanka. The Institute of Fundamental Studies, Kandy Sri Lanka

Vitousek PM, Aber JD, Howarth RW, Likens GE, Matson PA, Schindler DW, Schlesinger WH, Tilman DG (1997) Human alteration of the global nitrogen cycle: Sources and consequences. Ecological Applications 7: 737-750

Vogel H (2002) The soil nitrogen cycle. Report by the Environmental Geology: Division Department of Geological Survey (DGS) Lobatse Botswana, 25 pp

Wakode A, Vali SA, Deshmukh AN (1993) Fluorine content of drinking water and food stuff grown in an endemic fluorosis area Dongargaon District Chandrapur (MS). Gondawana Geol Mag 6: 1-15

Wang W, Li R, Tan J, Luo K, Yang L, Li H, Li Y (2002) Adsorption and leaching of fluoride in soils of China. Fluoride 35(2): 122-129

Wang Z (1991) Environmental behaviour and chemical transformation of Se in Chinese environments. PhD thesis of Research Centre for Eco-Environmental Sciences Chinese Academy of Sciences Beijing China

Wang Z, Gao Y (2001) Biogeochemical cycling of selenium in Chinese environments. Applied Geochemistry 16: 1345-1351

Warnakulasuriya KAAS, Balasuriya S, Perera PAJ, Peiris LCL (1992) Determining optimal levels of fluoride in drinking water hot dry climates- a case study in Sri Lanka. Community Dental Oral Epidemiology 2: 364-367

Warren HV, Delavault RE, Cross CH (1972) Possible correlations between geology and some disease patterns. Annals of New York, Acad Sci 136: 696

Wedepohl EH (1972) Handbook of Geochemistry. Springer, Berlin

Wedepohl KH (1996) The importance of the pioneering work by VM Goldschmidt for modern geochemistry (Plenary Lecture-Goldschmidt Conf Heidelberg, April 1996)

Weerasooriya SVR, Jinadasa KBPN, Dissanayake CB (1989) Decontamination of fluoride from community water supplies. Environ Tech letters 10: 23-28

Weerasooriya SVR, Priyadharshini KWV, Dissanayake CB (1994) Fluoride adsorption by goethite in iodide mediated simulated environmental systems. Toxicological and Environ Chemistry 44: 113-121

Wellenstein B, Wellenstein L (2000) Mineral nutrition for slipper orchid growers. An Tec Laboratory, USA (http://ladyslipper.com)

White JG, Zasoski RJ (1999) Mapping soil micronutrients. Field Crops Research 60: 11-26

Whitehead DC (1974) The influence of organic matter chalk and sesquioxides on the solubility of iodine elemental iodine and iodate incubated with soil. Euro J Soil Science 25: 461-470

Whitford GM, Pashley DH (1984) Fluoride absorption: The influence of gastric acidity. Calcified Tissues Int 36: 302-307

Wijewardane GG (2005) Biomineralization of urinary calculi (kidney stones): a geochemical study. University of Peradeniya, Unpublished B.Sc. thesis.
WHO (1971) International standards for drinking water. WHO Geneva
WHO (1978) How trace elements in water contribute to health. WHO Chronicle 32:381-385
WHO (1987) Technology for water supply and sanitation in developing countries. Report of a WHO Study Group, Geneva World Health Organ: Tech Rep Ser 7 742: 38 pp
WHO (1993) Guidelines for drinking water quality. 2nd volume 1, World Health Organization Geneva
WHO (1996) Trace elements in human nutrition and health. WHO Geneva
WHO (2001) Assessment of iodine deficiency disorders and monitoring their elimination. A guide for programme managers (2nd ed) ICCIDD, UNICEF WHO publication
WHO (2002) Environmental health criteria-225. Principles and methods for the assessment of risk from essential trace elements. WHO Publication Geneva
WHO/MDIS (1993) Micronutrient deficiency information system- global prevalence of iodine deficiency disorders. National Goitre Survey 1988, MDIS Working paper, Nr L WHO/UNICEF/ICCIDD
Wilkister K, Moturi N, Tole MP, Davies TC (2002) The contribution of drinking water towards dental fluorosis: A case study of Njoro Division, Nakuru District, Kenya. Environmental Geochemistry and Health 24: 123-130
Wilson DC (1954) Goitre in Ceylon and Nigeria. British J Nutr 8: 90-99
Wilson MJ (2003) Clay mineralogical and related characteristics of geophagic materials. J Chemical Ecology 29(7): 1525-1547
Woywodt A, Kiss A (1999) Perforation of the sigmoid colon due to geophagia. Archives of Surgery 134: 88-89
Wang W, Li R, Tan J, Luo K, Yang L, Li H, Li Y (2002) Absorption and leaching of fluoride in soils in China. Fluoride 35(2): 122-129
Xia WP, Tang JA (1990) Comparative studies for selenium contents in Chinese rocks. ACTA Scientiae Circumstantiae (China) 10: 125-132
Xie X, Cheng H (2001) Global geochemical mapping and its implementation in the Asia-Pacific region. Applied Geochemistry 16: 1309-1321
Xie X, Yin B (1993) Geochemical patterns from local to global. J Geochem Expl 47: 109-129
Xie X, Zheng K (1983) Recent advances in geochemical exploration in China. J Geochem Explor 19: 423-444
Xie X, Liu D, Xiang Y, Yunchuan X, Yan G, Lian C (2004) Geochemical blocks for predicting large ore deposits-concept and methodology. J Geochem Explor 84: 77-91
Xie X, Mu X, Ren T (1997) Geochemical mapping in China. J Geochem Explor 60: 99-113
Xie X, Sun H, Li S (1981) Geochemical exploration in China. J Geochem Explor 15: 489-506

Yang G, Xia YM (1995) Studies on human dietary requirement and safe range of dietary intake of selenium in China and their application to the prevention of related endemic diseases. Biomed Environ Sci 8: 187-201

Yang G, Peterson PJ, Williams WP, Wang W, Ribang L, Tan J (2003) Developing environmental health indicator as policy tools for endemic fluorosis management in the Peoples Republic of China. Environmental Geochemistry and Health 25: 281-295

Yoshizawa K, Willett WC, Morris SJ, Stampfer MJ, Spiegelman D, Rimm EB, Giovanucci E (1998) Study of prediagnostic selenium level in toenails and the risk of advanced prostate cancer. Journal of the National Cancer Institute 90: 1219-1224

Zha YR, Tao ZF, Wei LX (1996) Epidemiological survey in a high background radiation area Yangjiang China. (In Chinese) Zhonghua Liu Xing Bing Xue Za Zhi 17(6): 328-332

Zheng L (1991) Characterization of content and distribution of microelements in soils of China. In: Portch S (ed) International symposium on the role of sulphur magnesium and micronutrients in balanced plant nutrition. Potash and Phosphate Institute of Canada-Hong Kong-China, p 54-61

Zheng B, Hong Y (1988) Geochemical environment related to human endemic fluorosis in China. In: Thornton I (ed) Geochemistry and Health. Science Reviews Ltd Northwood, p 93-106

Ziauddin M, Roy S (1970) Geochemistry of nickel in the weathering cycle Sukinda area, Orissa (India). Seminar on Laterites India, p 194-197

Ziegler JL (1993) Endemic Kaposi's sarcoma in Africa and local volcanic soils. Lancet 342: 1348-1351

Zobrist J, Dowdle PR, Davis JA, Oremland RS (2000) Mobilization of arsenite by dissimilatory reduction of adsorbed arsenate. Environ Sci Technol 34: 4747-4753

Web Sources

http://www.unu.edu/env/arsenic/Dhaka
http://www.unizh.ch/amicroeco/uni/kurs/mikock/results/projectz/arsenihtml
http://www.healthyeatingclu.com/info/books/foodfacts

Subject Index

abiotic reactions, 171
acidic soil, 66, 68, 109, 134
aetiology, 3, 7, 28, 54, 125, 197, 233, 255, 256
Africa, 87, 100, 105, 135, 136, 144, 196, 231, 233,255
AIDS, 217
altitude, 22, 114, 115, 135, 231
ammonium, 110, 140, 146, 185, 200, 202, 203
anthropogenic sources, 67, 167
apatite, 60, 71, 72, 77, 84, 202-204, 257
aquatic systems, 111, 164
Argentina, 6, 157, 160
arsenate, 162, 165-174,183-186
arsenic in groundwater, 167, 181, 182, 185
arsenic, 6, 8, 13, 157-187, 190
arsenicosis, 187, 188, 189
arsenite, 162, 166, 169, 171, 172, 174, 183-186
arsenopyrite, 13, 173, 179, 183
assimilatory reduction, 212, 213
atherosclerosis 195, 216
atmosphere, 2, 6, 56, 101, 102, 122, 139, 141, 159, 257
bacteria, 112, 140, 141, 147-150, 152, 153, 204, 213, 215
Bangladesh, 6, 40, 99, 101, 142, 157, 159, 160, 163, 169, 171, 173, 175-185, 187, 225, 241, 253
baseline geochemical data, 251
bentonites, 61, 66, 67
bioaccumulation, 13, 47, 53, 211, 212
bioavailability, 28, 47- 51, 53, 54, 57, 58, 69, 107-109, 114, 119, 122, 128, 134, 159, 200, 206, 208, 221, 227, 251, 255
biogeochemistry, 1, 37, 171
biomethylation, 215
biomineralization, 204

biosphere, 2, 4, 56, 159, 171
bone, 17, 60, 69-71, 76-78, 96, 97, 193, 257
boron, 15, 16, 41
Botswana, 12, 150, 152
Brazil, 20, 24, 160, 198, 237-241, 247
brick tea, 94
British Geological Survey, 4, 9, 90, 113, 115, 146, 150, 157, 175, 176, 179, 181, 182, 256, 257
brushite, 203
calcium carbonate, 20, 66, 204, 244
calcium phosphate, 200, 202, 203, 239
calcium, 10, 14, 17, 20, 41, 59, 63, 66, 71, 77, 91, 114, 146, 192, 193, 196, 200, 202, 204, 239, 244
Cambodia, 157
cancer, 6, 7, 17, 18, 56, 69, 109, 139, 142, 153, 154, 155, 189
carbonate apatite, 200, 202, 203, 239
carcinogenicity, 154
cardiomyopathy, 197, 216
cardiovascular diseases, 8, 40, 191, 192, 196, 218, 256
caries, 18, 69, 74, 84
cassava, 121, 122, 136
cation exchange, 37, 47, 63, 140, 225, 228
causation, 55, 155, 218
cerium oxide, 238, 239
chemical speciation, 4, 255
chemical weathering, 24, 30, 33
Chiba prefecture, 115
Chile, 157, 160
China, 5, 5, 59, 92-94, 132, 147, 154, 157, 205, 219-221, 226, 235, 237, 238, 241, 245, 247, 251-257
chloroapatite 71
clays, 6, 21, 25, 26, 32, 52, 67, 109, 168, 169

coal, 13, 73, 92-94, 97, 106, 161, 163, 206, 207, 220, 225
coastal sands, 198, 238
colloids, 140
community fluorosis index, 74, 75
contaminant, 40, 48, 49, 51, 53, 165
continental crust, 103, 104
copper, 14-16, 18, 41, 53, 57, 114, 227, 255
coronary heart diseases, 154, 192, 196
correlation coefficient, 191, 192, 193
cretinism, 99, 120, 136
cystine, 200, 216
deep well, 64, 65, 81, 84, 95, 126, 157, 152
defluoridation, 95-97
deltaic sediments, 157, 160
denitrification, 140, 141
dental fluorosis, 5, 59, 60, 69, 71, 73, 74, 78, 81, 84
dentine, 70
dimethylselenide, 208, 215, 216
dissimilatory reduction, 172, 173, 212
dose-response relationship, 55, 192
dug well, 64. 65, 81, 146, 180
duricrust, 31
East African rift valley, 87, 88
ecosystems, 140, 141, 142
Eh-pH, 108, 109, 164, 166
enamel, 61, 69-75
endemic goitre, 5, 100, 122, 123, 125, 126, 128, 130, 135, 136, 252
endomyocardial fibrosis, 197, 199
enzymes, 14, 111, 209, 210
epidemiology, 44, 54, 55, 154, 205, 251, 256
essential trace elements, 9, 24, 26
Ethiopia, 7, 91, 92, 231, 232
eukaryotes, 111
eutrophication, 141
faecal pollution, 149
ferralsols, 21-23
fertilizers, 7, 61, 66, 68, 141-144, 146, 147, 152, 183, 184, 200
fluorapatite, 60, 71
fluoride, 5, 10, 13, 16, 17, 41, 59-64, 66-71, 73-96, 204, 252, 254
fluoride-rich minerals, 63
fluorine, 10, 61, 62, 66, 67, 71, 79, 81, 83, 84, 88, 93
fluorite, 61, 62, 85

gamma rays, 247
geochemical blocks, 252
geochemical cycles, 2-4, 56, 169, 235
geochemical diseases, 233, 252
geochemical maps, 251, 252, 256
geochemical patterns, 252-254, 256
geophagy, 223-230
Ghana, 5, 73, 89, 90, 160
goethite, 25, 29, 168
goiter belt, 120, 132, 137
Goldschmidt VM, 1, 2
Guarapari, 237, 238, 239, 241, 245, 247
halides, 60, 62
halloysite, 224, 226, 228
hard water, 71, 191, 192, 200, 204
hardness, 8, 17, 40, 71, 127, 191, 192, 193, 194, 250, 254
hematite, 25, 29, 34, 123, 165
heterotrophic bacteria, 141
high background radiation areas, 237, 238, 240, 243, 245, 247
Highland Complex, 79, 81
homeostasis, 12, 56, 57, 70, 71
hormones, 10, 117, 118, 120, 123, 136
human wastes, 152
humic substances, 37, 109, 123
hydrogeochemical atlas, 81, 252
hydrogeochemistry, 35, 59
hydrosphere, 2, 56, 141
hydroxyapatite, 71
hyperaccumulation, 210
hyperkalemia, 230
hypertension, 193, 195, 216
hypothyroidism, 118, 120, 136, 218
ilmenite, 165, 178, 238, 239, 240
India, 5, 6, 12, 17, 31, 59, 62, 73, 84-86, 96, 99, 101, 130-132, 142, 148, 157, 160, 187, 197, 199, 200, 204, 224, 235, 237, 238, 240-242, 245, 247, 249, 250
infertile soils, 27
iodate, 102, 109
iodide, 41, 102, 107, 109, 111, 113, 117, 121, 122, 126, 135, 136
iodine cycling, 114, 115
iodine deficiency disorders (IDD), 3, 5, 99, 101, 107, 108, 114, 115, 117, 118, 120, 122, 123, 125-128, 132, 135-137, 191, 193, 194, 204
iodine fixation potential, 107

Subject Index

iodine, 3, 5, 6, 59, 99-103, 105, 107-137, 218, 222, 257
Iran, 154, 237, 238, 241, 243, 244, 247, 248
Iron oxides, 6, 107, 168, 169, 179, 184,
Iron, 10, 14, 15, 18, 30, 31, 40, 41, 51-53, 70, 95, 114, 163, 170, 173, 178, 183, 192, 206, 227, 229, 230, 232, 239, 256
Kaposi's Sarcoma, 233, 234
Kashin-Beck disease, 219, 221
Kenya, 7, 73, 84, 87, 88, 89, 135, 226, 231
Kerala, 85, 132, 197-199, 237, 238, 240, 241, 245, 247, 249, 250
keratosis, 187, 188
Keshan disease, 205, 216, 221, 252
lanthanum oxide, 238
laterite, 25, 26, 28-30, 32-24, 101, 107, 221
light rare earth elements, 239
lithosphere, 2, 56
magnesium, 14, 15, 17, 41, 71, 192, 193, 195, 196, 200, 2002, 203
Maldives, 130, 132
malnutrition, 119
manganese, 14-16, 18, 41, 51, 53
medical geochemistry, 3, 4, 6, 7, 25, 39, 40, 43, 123, 139, 159, 251, 256, 257
medical geology, 3, 4, 5, 7, 12, 15, 19, 28, 35, 40, 47, 54-56, 59, 99, 107, 114, 125, 139, 157, 175, 205, 219, 223, 235, 251, 252
Meghna, 175, 176, 182
melanosis, 187, 188
mental deficiency, 118, 119, 120
metal speciation, 47, 49
metamorphic rocks, 79, 102, 103, 161, 163, 175
methaemoglobinaemia,7, 149, 153
methylation, 57, 211, 212, 215, 216
mica, 66, 67, 86, 102, 103
microbial activity, 111, 167, 173
microbial transformation, 211, 212
microorganisms, 111, 140, 147, 169, 170, 171, 174, 212-214, 216, 224, 245
mineral deficiencies, 7, 10
mineralized terrains, 29
Minjingu phosphate mine, 243

molybdenum, 13, 15, 16, 41, 109
monazite, 198, 199, 237-242, 245, 250
montmorillonite, 66, 169, 170, 179
Morocco, 114, 115, 152
Myanmar, 99, 101, 115, 157
myocardial infarction, 193, 195, 196
nanobacteria, 204
natural dust, 234,235
natural radiation, 237, 238, 240, 241, 243, 244, 247-250, 257
natural reactor, 245, 246
neodymium, 238, 245
Nepal, 99, 101, 130, 157
nephrolithiasis, 204
newberyite, 203
nickel, 29, 32, 33
Nigeria, 137, 197
nitrate pollution, 12, 141, 144, 145, 150, 152
nitrates, 7, 12, 40, 139, 14, 142, 148-155, 171
nitrobacter, 140
nitrogen cycle, 139-141
nitrogen transformations, 139, 141
nitrogen, 7, 14, 15, 40, 139-142, 144-147, 152, 153, 155
nitrogenous fertilizer, 139, 146
oceanic crust, 104
oesophageal cancer, 7, 154
Oklo, 245, 246
organic matter, 27, 37, 39, 67, 71, 101, 104, 107, 110, 122, 123, 128, 152, 169, 172, 179, 184, 206, 208, 221, 225
organic nitrogen, 140, 152
Oryza sativa, 147, 185
osteomalacia, 71
osteoporosis, 193
Ounein valley, 114, 115
oxalate, 200, 202-204
oxyanions, 165, 211, 212, 213, 215
Pakistan, 130, 142, 153
Paracelsus, 9
pathogenicity, 233
pathogens, 148
peat, 81, 105, 107, 157, 161, 164, 169, 172, 173, 175, 178-180, 185
phosphates, 183, 184
phosphorus, 13, 14, 184
Piper plots, 40
pit latrines, 148-152

pituitary gland, 117
plate tectonics, 2, 114, 115
pneumoconiosis, 234
podoconiosis, 7, 223, 231-134
potassium, 9, 10, 14, 15, 146, 230
Precambrian, 79
prokaryotic, 171, 172
radioactive iodine, 111
radioactivity, 198, 237, 238, 240-244, 250
radioisotope/isotope 101, 110, 245
radium, 41, 239, 240, 244
radon, 237
Rajasthan, 85
Ramsar, 237, 238, 241, 243, 244, 247-249
recommended dietary allowance, 112
relative mobility, 36, 29, 51, 52
rice, 69, 111, 112, 114, 121, 126, 129, 130, 131, 147, 185, 186, 200, 256
risk assessment, 51-53, 57
risk factor, 54, 55
rock weathering, 21, 24, 206, 231
rutile, 24, 238
sanitation, 12, 148, 151, 152
scavengers, 28, 29
secondary minerals, 32
sedimentary rocks, 66, 67, 81, 104, 105, 161, 163, 206, 244
selenate, 208-211, 213
selenite, 208, 209, 211, 215
selenium cycle, 212
selenium, 13, 16, 18, 41, 109, 114, 123, 126, 129, 136, 205-222, 252
selenocysteine, 213, 216
Senegal, 84, 151
Se-oxyanions, 211, 212, 213, 215
serpentinites, 30
shales, 13, 66, 104, 161, 206
silicates, 193, 228
silt, 107, 163, 178, 226
skeletal fluorosis, 3, 5, 44, 59, 60, 69, 73, 74, 76, 77, 78, 84, 91-94
smectite, 21, 25, 28, 29, 224, 225, 226, 227, 228
soakaways, 149, 150
sodium, 9, 41, 71, 146, 227
soft water, 191, 195, 204
soil environment, 51, 111
soil solution, 47, 49, 185

soil-plant system, 111
solubility 37, 52, 141, 164, 165, 198, 233
sorption, 38, 51, 52, 67, 107, 109, 111, 114, 123, 124, 169, 170, 174
South Africa, 105, 196, 256
Sri Lanka, 5, 12, 26, 30-32, 73, 74, 78-84, 90, 96, 99, 100, 101, 125,-129, 146, 150, 151, 222, 227, 252
struvite, 200, 202, 203
subduction, 115, 116
sulphide minerals, 28, 161, 167, 206
sulphides, 28, 161, 167, 206
sulphur, 161, 179, 205, 206, 210, 212, 213, 215
Taiwan, 6, 92, 157, 160, 187
Tanzania, 5, 64, 90, 91, 145, 228, 243
tea, 68, 92, 94, 97
teeth, 17, 18, 60, 61, 70, 71, 73, 193, 257
thermal springs, 90, 160
thorium, 41, 238-240
thyroid stimulating hormone, 117
thyroid, 18, 117-123, 136, 202, 218, 222
thyroxine, 117
toxicity, 7, 36, 50, 51, 57, 59, 69, 153, 171, 174, 209-223, 216, 218, 221, 229, 232, 253, 255, 256
tropical countries, 73, 76, 84, 95, 197, 200, 255
tropical environment, 19, 20, 24-29, 35-37, 40-44, 54, 99, 101, 107, 251
Uganda, 135, 199, 225
ultrabasic rocks, 30, 34, 163, 219
uranium oxide, 238, 245
uranium, 41, 184, 243
urea, 146, 147, 202, 203
uric acid, 200, 202
urinary stones, 191, 200, 203
urolithiasis, 204
Vietnam, 130, 132, 133, 253
Vijayan Complex, 79, 83, 84
volatilization, 111, 112, 212
volcano-clastics, 66, 81
Wanni Complex, 79
water factor, 192, 194
water hardness, 8, 40, 191-193, 200, 204
water table, 30, 87, 171, 183
water-rock interaction, 64, 85

Subject Index

West Bengal, 6, 85, 157, 160, 169, 175, 178, 187, 189, 190
World Health Organization (WHO), 11, 12, 16, 51, 50, 57, 60, 69, 73, 74, 101, 117-119, 121, 135, 137, 148, 151, 157, 181, 195, 208, 209

X-ray, 4, 33, 78, 203, 230, 231, 235, 247
Yangjiang, 237, 241, 245, 247, 248
Zaire, 120, 136
Zimbabwe, 137, 226
zinc, 13, 14, 15, 18, 41, 53, 114, 230, 255